BN 9408749 0
AF843830

LOGICAL SEMIOTICS AND MEREOLOGY

FOUNDATIONS OF SEMIOTICS

General Editor
ACHIM ESCHBACH
(University of Essen)

Advisory Editorial Board

Herbert E. Brekle (Regensburg); Geoffrey L. Bursill-Hall (Burnaby, B.C.)
Eugenio Coseriu (Tübingen); Marcelo Dascal (Tel-Aviv)
Lambertus M. de Rijk (Leiden); Max H. Fisch (Indianapolis)
Rudolf Haller (Graz); Robert E. Innis (Lowell, Mass.)
Norman Kretzmann (Ithaca, N.Y.); W. Keith Percival (Lawrence, Kansas)
Jan Sulowski (Warszawa); Jürgen Trabant (Berlin)

Volume 16

Richard M. Martin

Logical Semiotics and Mereology

LOGICAL SEMIOTICS AND MEREOLOGY

RICHARD M. MARTIN

Withdrawn from
University of Brighton
Libraries

JOHN BENJAMINS PUBLISHING COMPANY
AMSTERDAM/PHILADELPHIA

1992

UNIVERSITY
LEARNING
24 NOV 1994
RESOURCES
OF BRIGHTON

401.41
MAR

Library of Congress Cataloging-in-Publication Data

Martin, Richard M.
Logical semiotics and mereology / Richard M. Martin.
p. cm. -- (Foundations of semiotics, ISSN 0168-2555; v. 16)
Includes bibliographical references.
1. Semiotics. 2. Language and languages--Philosophy. 3. Logic. 4. Whole and parts (Philosophy) 5. Science--Philosophy. I. Title. II. Series.
P99.M35 1992
302.2 -- dc20 91-41368
ISBN 90 272 3288 1 (alk. paper) CIP

9408749

© Copyright 1992 - John Benjamins B.V.
No part of this book may be reproduced in any form, by print, photoprint, microfilm, or any other means, without written permission from the publisher.

John Benjamins Publishing Co. · P.O. Box 75577 · 1070 AN Amsterdam · The Netherlands
John Benjamins North America · 821 Bethlehem Pike · Philadelphia, PA 19118 · USA

Other books by the same author

Truth and Denotation
The Notion of Analytic Truth
Toward a Systematic Pragmatics
Intension and Decision
Belief, Existence, and Meaning
Logic, Language, and Metaphysics
Whitehead's Categoreal Scheme and Other Papers
Events, Reference, and Logical Form
Semiotics and Linguistic Structure
Peirce's Logic of Relations and Other Studies
Pragmatics, Truth, and Language
Primordiality, Science, and Value
Logico-Linguistic Papers
Mind, Modality, Meaning, and Method
Metaphysical Foundations: Mereology and Metalogic

To the memory of
Haskell B. Curry,
with admiration and gratitude

"Was vermied' ich denn die Wege,
Wo die andern Wand'rer geh'n,
Suche mir versteckte Stege
Durch verschneite Felsenhöh'n?
..
Einen Weiser seh' ich stehen
Unverrückt vor meinem Blick;
Eine Straße muß ich gehen,
Die noch keiner ging zurück."

W. Müller

Contents

Preface

The various papers collected together in this volume are concerned with a variety of vitally important topics in philosophic logic, the philosophy of language, the philosophy of mathematics, the philosophy of science, and the application of modern logic to wider philosophical problems. All of them make fundamental use, in one way or another, of logical semiotics, the modern trivium of systematic syntax, semantics, and pragmatics, and some of them, of mereology, the general theory of parts and wholes. In syntax the concern is with signs or expressions as such, in semantics, with how signs are related to their objects, and in pragmatics, with how signs are used in both linguistic and extra-linguistic contexts. In recent years, mereology has been showing itself to be of much greater importance than has heretofore been thought, and interesting new advances are being made in the subject. The papers here seek to carry forward the systematic development of these various subjects, to show how intimately they are interrelated, and to illustrate their utility in specific applications in linguistics, mathematics, science, and philosophy generally.

Very roughly, the papers here may be described as follows. In I, many uses of *negation* in natural language are distinguished in such a way as to exhibit its extraordinary "positive" power. II contains a sketch of a *logistic grammar* of natural language as based on a *process-* or *event-logic*. Central to this is a systematic pragmatics, or logial semiotics, the historical origins of which are discussed in III, especially in relation to the seminal contributions of *Carnap*. In IV, P.T. Geach's recent work on *comparatives* in natural language is subjected to a careful scrutiny, and reconstruction, in the light of the material in II and III.

In V, the author's pervious work on extended mereology, or the theory of part to whole, is given another interpretation along the lines of a *homogeneous logic of individuals*, somewhat in the manner of Leśniewski's *Ontologie*. In VI, the material of V is interrelated with the author's *non-translational semantics*, a kind of semantics differing radically from the usual formulations in containing no translation of the object-language part. In VII, still another interpretation of essentially the material of V and VI is given, a *Platonic* one in a certain sense. In V-VII fundamental use is made of the so-called *ordinal* entities, to provide for a full logic of relations and therewith the foundations of mathematics. In IX and X, the nature of ordinal entities is discussed in connection with the relevant work of *Peirce* and *Russell*, and as contrasted with that of *Goodman, Rolf Eberle*, and *Hempel*.

X is devoted to a critical discussion of *Hao Wang's* views concerning the notion of *set*. In XI, *Patrick Suppes's* views concerning *pluralism* and the "*mainstream*" of science are subjected to a careful scrutiny, as are the views of *Philip Davis* and *Reuben Hersh* concerning "mathematical experience," in XII.

In XIII, it is shown how a *quasi-combinatorial* logic of *relational terms* can be formulated, once we become clear about the inner structure of the algebra of *n*-adic relations. XIV is devoted to a critical analysis of *Husserl's* theory of wholes and parts as contained in his *Logische Untersuchungen*. Further discussion of wholes and parts, and the role of "compound" individuals in the languages of science, is given in XV. XVI consists of a critical discussion of *Paul Weiss's* early logic papers, and XVII, of *Quine's* recent "Facts of the Matter" as contrasted with the views of the eminent physicist *J.A. Wheeler*.

In XVIII an extensive theory of *mereological sums* is put forward with a view to showing how the *quantifiers* may be handled within it, and how the "*collective*" *use of terms* in natural language may be handled. The "mystery" of the effectiveness of mathematics in the natural sciences is discussed in XIX, centering around the views of *Wigner, Whitehead*, and *Carnap*. Finally, in XX, a new approach to the foundations of mathematics is put forward, based on *ordinal acts* rather than upon ordinal entities. In this view, mathematics may be seen as of a piece — both ontologically and structurally — with physics and the rest of science, rather than as a subject apart in which some new realm of entities such as "sets" must be recognized.

The various papers here are to some extent independent of each other and for the most part may be read separately. Even so, they do, of course, help to reinforce one another, in detail as well as in general intent.

The author wishes to thank the Boston University Center for the Philosophy and History of Science, Northwestern University, and the Nitram Foundation for support during the writing of these papers. Thanks are due also the editors and publishers of the journal *Krisis*, the *Journal of Chinese Philosophy*, the series *Pragmatik*, and the journal *Semiosis*, respectively, for kind permission to use here in a revised form the papers I, II, III, and XIII, originally published or to be published by them.

CHAPTER ONE

On the positive power of negation

> "Si antiquitatem spectes, est vertustissima, si dignitatem est honoriatissim, si juristicdionem est capacissima."

The little word 'not', generally used to express negation, is one of the mighty, three-letter monosyllables in English. The logical analysis of it and its variants, as occuring in all possible contexts of use, has scarcely begun in a serious way. Professional linguists tend to have stayed clear of such analysis, on the grounds that it would become too "dangerously adulterated with formal logic."[1] In its place linguists have tended to fall back on traditional methods in grammar, which is to say on "impressionistic, non-rigorous" methods consisting mostly of illogicalities, incorrrect or at best vague modes of reasoning, in short, on horse-and-buggy, rough and ready techniques in this day and age of highly sophisticated methodologies. That this is the case is a near-scandal in linguistics, in linguistic philosophy, as well as in the newer "philosophical logic."[2]

In this paper let us sketch the outlines of a broad logico-semiotical system in which '~' for classical — as opposed to intuitionistic or some other — negation plays the fundamental role. From a formal point of view, nothing strikingly new will be attempted here, but it will be of interest to pull together various strands of work that have been done elsewhere — threaded together, it is hoped, into a rope, or at least a cord, of some strength.

Given a sentence 'A', ~ 'A' is to express the negation of 'A'. Later we will observe how negation in this sense is intimately connected with the semantical notions of the truth and falsity of sentences. Even before doing this, however, we may note that the little sign '~', all by itself, has a remarkable role to play, so fundamental a one in fact that any serious logico-metaphysical or logico-semiotical view seems unthinkable without it or some surrogate for it. '∼' together with the other familiar truth-functional connectives 'v' (corresponding roughly with some basic uses of 'or' in English), '.' (roughly for some basic uses of 'and'[3]), '⊃' (very roughly for 'only if'), and '≡' (for 'if and only if'), have been well-characterized in truth-functional theory, perhaps the most widely accepted part of modern logic. These several notions "interanimate" each other so intimately that one without the others is barren and of no avail.

The full power of '∼' comes to light best by considering successively various extensions of the basic truth-functional theory. Let us go on now, step by step, to reflect on the most important of these, in the sense of being relevant to the general aim of gaining an appropriate logico-semiotical framework. The first extension is of course into the theory of quantification. Let us write '(*x*)' as the universal quantifier, or '(∀*x*)' or even '(A*x*)' if you will. Then

(1) '(*x*)-*x*--'

is to express that all *x* are such that --*x*--, where '--*x*--' is a sentential form of the language. Immediately a new context arises in which '∼' may occur. Thus

(2) '∼ (*x*)--*x*--'

expresses that it is *not* the case that for all *x*, --*x*--. Another context is then

(3) '(*x*) ∼ --*x*--',

expressing that for all *x* it is not the case that --*x*--. And still another context is

(4) '∼ (*x*) ∼ --*x*--',

expressing that not all *x* are such that ∼ --*x*--. This latter will be recognized as expressing that there is something, perhaps even the null thing,[4] such that --*x*--, and this is commonly expressed by

(5) '(E*x*)--*x*--'.

We have then some additional contexts such as

(6) '∼ (E*x*)--x--',

equivalent to (3),

(7) '(E*x*) ∼ --*x*--',

equivalent to (2), and

(8) '$\sim$ (Ex) $\sim$ --x--',

equivalent to (1) itself.

What now is a "sentential form" of the language? it might be asked. Well, the language is presumed to contain suitable non-logical primitive predicates, each of appropriate degree, in terms of which the subject-matter of the language may be handled. Thus 'Per' might be a primitive one-place predicate for human persons, so that 'Per x' would express that x is a human person. 'Older' could be a two-place relational predicate for being older than, significant in contexts of the form 'x Older y'.

'(Per x . Per y . x Older y)'

would express that person x is older than person y. And so on, for relational predicates of higher degree of or a greater number of places. And so on, also for all manner of different subject-matters no matter how intricate or complex.

Suppose 'F_o' is primitive one-place predicate.

(9) '$\sim F_o x$'

then express that x is not an F_o. (But '$F_o \sim x$' and '$F_o\, x \sim$' are of course meaningless.) Similarly for

(10) '$\sim x\, R_o\, y$',

where 'R_o' is a primitive two-place predicate, expressing that x does not bear R_o to y (where '$x \sim R_o\, y$', '$x\, R_o \sim y$', and '$(\sim x)\, R_o\, y$' are meaningless).

(9) and (10) are of special interest in that they give us a new kind of negation. We may let

(11) '$(\text{-}F_o)\, x$' abbreviate '$\sim F_o\, x$'

and

(12) '$x\, (\text{-}R_o)\, y$' abbreviate '$\sim x\, R_o\, y$'

thus gaining in context the notion of the negation or negative of F_o and of the negation or negative of R_o.

These new kinds of negatives may readily be generalized to all *definable* predicates of the system. A convenient way of expressing such definable predicates is by means of the notation for *virtual* classes and relations.[5]

Let

> '$\{x \ni$ --*x*--$\}$ *y*' be an alternative notation for '--*y*--', where '--*y*--' differs from '--*x*--' only in containing free occurrences of '*y*' wherever there are free occurrences of '*x*' in '--*x*--'.

'$\{x \ni$ --*x*--$\}$' is a one-place *abstract* and the context '$\{x \ni$ --*x*--$\}$ *y*' states that *y* is a *member* of the virtual class of all *x*'s such that --*x*--.

Let *F* and *G* now be any virtual classes. We may then define the *logical sums* and *products* of them as follows.

> '$(F \cup G)$' abbreviates '$\{x \ni (F x \vee G x)\}$',
> '$(F \cap G)$' abbreviates '$\{x \ni (F x \,.\, G x)\}$'

And (10) may be generalized into the following, where *F* is any virtual class.

> '$-F$' abbreviates '$\{x \ni \sim F x\}$'.

And of course the *universal* and *null virtual classes* are then also immediately forthcoming by letting

> 'V' abbreviate '$\{x \ni (\mathrm{F}_0 x \vee \sim \mathrm{F}_0 x)\}$',

where 'F_0' is a primitive predicate, and

> 'Λ' abbreviate '-V'.

These notions will be recognized as adaptations of the familiar ones of a Boolean algebra.

In a similar way we can let

> '$\{xy \ni$ --*x*--*y*--$\}$ *zw*' be an alternative writing of '--*z*--*w*--', where (etc. as required),

in order to gain a notation for saying that *z* stands in the dyadic virtual relation $\{xy \ni$ --*x*--*y*--$\}$ to *w*. We can then introduce logical sums, products, and negatives of dyadic virtual relations, essentially as above *mutatis mutandis*, and negatives and the universal and null dyadic virtual relations. And similarly for triadic virtual relations, and so on.

It is interesting to note that, if '$\sim$' and 'v' (for disjunction) are taken as the truth-functional primitives of the theory, then in familiar fashion we may define, using Quine's corner-quotes,

> $\ulcorner(A \,.\, B)\urcorner$ as $\ulcorner \sim (\sim A \vee \sim B)\urcorner$.

Thus even conjunction, as construed in this way, is defined in terms of the basic negation. Usually the '.' for 'and' is regarded as a "positive" notion in some sense. However, as will be evident throughout, the '$\sim$' intertwines so intimately with other notions, and in such deep and basic ways, that notions we would usually dub as 'positive' are seen no longer to be so in any genuine sense. Also there are always double negations of sentences, which are both positive and negative, albeit in different senses. *Duplex negatio affirmat.*

The theory of *n*-adic virtual relations, for $n \geqq 2$, contains a motley of important concepts of the highest interest for philosophy. As an example, consider the *converse* of a dyadic virtual relation, defined by

'$\cup R$' for '$\{xy \ni R\, yx\}$',

or the *relative product* of two dyadic virtual relations,

'(R/S)', defined as '$\{xy \ni (\mathrm{E}z)\,(R\, xz \,.\, Szy)\}$'.

The converse of the relation of being-older-than is that of being-younger-than. The relative product of the relation of being-the-mother-of and of being-the-father-of is the relation of being-the-paternal-grandmother-of. Bear in mind that we also have converses of negatives, relative products of negatives, and negatives of a host of other notions that constitutes the theory of dyadic virtual relations. And similarly for triadic virtual relations, and so on.[6] Already we see that we have arrived at a commodious infinity of uses of '~'. And even so, we have scarcely begun!

The next extension to consider is into the so-called calculus of individuals, essentially Leśniewski's *mereology* or theory of the part-whole relation between individuals. Let 'x P y' express primitively that the individual x is a part of the individual y, in perhaps some spatio-temporal sense, although other interpretations are of interest also. In terms of 'P', logical sums and products of individuals may be introduced, as well as negatives, a universal or world-individual, and even the null one. This latter is essentially Leśniewski's "one individual that does not exist," but should not be confused with the null class Λ. According to this theory, then, given any two individuals x and y, there exists a sum-individual consisting of x together with y, no matter how disparate or scattered they may be. And similarly a product, which may of course be null if the two individuals have no common spatio-temporal part — on the spatio-temporal interpretation of 'P'. The negative of an individual is then the rest of the world taken together as a whole. Thus the sum of an individual with its negative is the world-individual. If x is an ordinary object of daily experience, think how vast an entity its negative is![7]

In terms of 'P', identity of individuals may be immediately introduced, by letting

'$x = y$' be short for '(x P y . y P x)',

or, in primitive notation, as short for

'~ (~ x P y v ~ y P x)'.

Note that, in view of this definition, identity, a most "positive" notion surely, turns out to be a negative one, even a triple negative one in a certain sense.

Heinrich Scholz, in his remarkable little book *Metaphysik als Strenge Wissenschaft*[8], regarded identity as the first notion we need the moment we leave logic to enter the portals of metaphysics proper. In view of the above definition of '=', however, it would seem more appropriate to regard a relation of part-to-whole as carrying this honor. The theory of part-to-whole is perhaps no longer a theory of logic, but immediately one of metaphysics, embroiling us in deep problems concerning the inner structure of whatever things or entities are taken as individuals. Traditionally it is the study of ontology that has dealt with such problems. Logico-metaphysics, as conceived here, is essentially traditional ontology in modern grab.

Thus far negation has been considered only at the object-language level. Modern logic and metaphysics, however, are unthinkable without an extensive development of *meta*logic. Metalogic, or logical semiotics, divides conveniently into syntax, semantics, and pragmatics. Although the systematic development of metalogic has taken place only in recent years, there have been important anticipations of it throughout the long history of logic, especially during the middle ages.[9] Let us reflect for a moment upon a few of the great multiplicity of fundamental metalogical locutions.

The basic notion of syntax is that of *concatenation*. Let a and b be any expressions of the object-language. Their concatenate ($a^\frown b$) is then merely a followed by b in the normal way in which letters or words follow one another on the printed page, or spoken phonemes follow one another in vocal speech or communication.[10] Concatenation-theory might be regarded as an extension of mereology. It seems better, however, for present purposes, to regard it as a new subject on its own, subject to its own vocabulary and laws. Thus we do not assign a negative to each expression of the language, and there is no need to introduce logical sums and products of two expressions. Nor need there be a null expression nor a universal one. Even without these, the subject is one of remarkable power, enabling us to portray the purely structural features of our language, i.e., features of language depending only upon the expressions and how they are interrelated with one another. The role of '∼' in syntax is thus essentially what it is in the theory of quantification with given non-logical primitive predicates. We can handle concatenation as a triadic relation C so that '*a* C *b*, *c*' may express that the expression *a* is the concatenate of *b* with *c*. Using a Russellian description we may then define

'$(a^\frown b)$', as used above, as '$(\iota c\,.\, c \,\mathrm{C}\, a,b)$',

so that $(a^\frown b)$ in the one and only expression that bears C to *a* and *b* in that order.

Strictly we should distinguish here two kinds of concatenation, depending upon whether the expressions of the language be taken as shapes or sign-designs or as inscriptions or sign-events. Two very different kinds of syntax emerge. Both are needed for a full theoretical syntax, although for many purposes a syntax of one or the other kind may be made to suffice.

Let us turn now for a moment to denotational semantics, the study of how language relates to the individuals or "things" or entities about which it speaks. Here 'Den' for denotation is taken as the basic notion, so that '*a* Den *x*' may express that the one-place predicate of the object-language denotes the object *x*.[11] This notion may be taken as the fundamental one of semantics, just as 'C' is for syntax. Here again the role of '∼' is primarily truth-functional and does not lead to essentially new material.

The most important definable notion in denotational semantics no doubt is that of truth. Let *a* be a sentence of the object-language, i.e., a sentential form containing no free variables. More particularly let *a* be '-----'. Under what semantical conditions is '-----' true? Just where some one-place abstract '$\{x \ni$ -----$\}$' bears Den to all objects. Let us consider the general case. Let 'Ex *b*' express that *b* is the variable '*x*'. Then we may let

'Tr *a*' abbreviate '$(\mathrm{Sent}\, a\,.\,(\mathrm{E}b)(\mathrm{Ex}\, b\,.\,(y)\ulcorner\{x \ni a\}\urcorner \mathrm{Den}\, y))$'.

a is thus true just where *a* is a sentence (Sent) and there exists a variable '*x*' such that $\ulcorner\{x \ni a\}\urcorner$ denotes all objects. This definition seems deceptively simple, but a good deal of rather sophisticated theory is needed to place it in its proper *ambiente*.[12]

Note that '∼' may occur now in '∼ Tr *a*' as well as in $\ulcorner\sim a\urcorner$ in the context 'Tr $\ulcorner \sim a \urcorner$'. It so happens that these contexts are equivalent, that

$$(a)(\mathrm{Sent}\, a \supset (\mathrm{Tr}\,\ulcorner\sim a\urcorner \equiv \sim \mathrm{Tr}\, a)),$$

in view of fundamental semantical laws. In particular we have a semantical law of excluded middle, that every sentence is either true or its negation is,

$$(a)\,(\mathrm{Sent}\ a \supset (\mathrm{Tr}\ a \vee \mathrm{Tr}\ \ulcorner\sim a\urcorner)),$$

and of course that

$$(a)((\mathrm{Sent}\ a\ .\ \mathrm{Tr}\ a) \supset \sim \mathrm{Tr}\ \ulcorner\sim a\urcorner),$$

that no negation of a true sentence is true. Falsehood is merely the negative of truth, so that

'Fls *a*' may abbreviate '(Sent *a* . ~ Tr *a*)'.

All of this is in accord with the classical conception of truth, and we must be careful not to conflate truth with any of its illicit surrogates.[13]

When we turn to pragmatics we find that negation plays a much more exciting rôle than in either syntax or semantics. Pragmatics is taken here, following Carnap in essentials, to be the general area of linguistic study in which the *user* of language is brought in fundamentally. (See III below.) As a key notion let us consider that of *acceptance*. Let

'*p* Acpt *a*, *e*'

express that person *p* accepts the sentence *a* on occasion *e*, i.e., takes it to be true, or is disposed to take it as true if asked, is willing to act as if it were true, and so on. Acceptance is not to be confuted with truth, as it is frequently. The notions are very different.[14] In particular note that we do not have a law of excluded middle for Acpt. It does *not* in general hold that

$$(p)(a)(e)(\mathrm{Sent}\ a \supset (p\ \mathrm{Acpt}\ a,e \vee p\ \mathrm{Acpt}\ \ulcorner\sim a\urcorner,e)).$$

Certainly not. We might have a sentence *a* such that *p* neither accepts it, on occasion *e*, nor its negation. We might then say that *a* is indeterminate for *p* at *e*. We could let

'*a* Indet *p*,*e*' abbreviate '(Sent *a* . ~ *p* Acpt *a*,*e* . ~ *p* Acpt ⌜~ *a*⌝,*e*)'.

Or *p* might *reject a* in the sense of not accepting *a* at e. Let us call this 'rejection$_1$'. Then

'*p* Rjct$_1$ *a*,e' abbreviates '(Sent *a* . ~ *p* Acpt *a*,*e*)'

as contrasted with rejection$_2$, of accepting the negation of a sentence, where

'*p* Rjct$_2$ *a*,*e*' abbreviates '(Sent *a* . *p* Acpt ⌜~ *a*⌝,*e*)'.

Other pragmatical relations, for *apprehending*, for *asserting*, and so on, may be handled in similar vein and in which many further interesting uses of negation emerge.[15]

We have skipped from denotational semantics to pragmatics with no mention of an intervening semantics of intensions. Traditionally intensions are supposed to capture the "meanings" of terms as over and against their designata or denotata. Of course 'meaning' is highly ambiguous. One way of construing it technically might be as follows. Let

'p $\text{Under}_e\ x,a$'

express that person p in the cognitive act e takes, or is disposed to take, object x as falling under the one-place predicate (PredConOne) a. The use of 'under' here is based on Frege's *Art des Gegebenseins* or mode of presentation[16] — construed here as a mode of *linguistic* presentation. Here again we must distinguish between

'$\sim p\ \text{Under}_e\ x,a$' and '$p\ \text{Under}_e\ x,\ulcorner\text{-}a\urcorner$'.

It should generally hold that

$(p)\ (e)\ (x)\ (a)\ (p\ \text{Under}_e\ x,\ a \supset \sim p\ \text{Under}_e\ x,\ \ulcorner\text{-}a\urcorner)$,

but *not* of course in general that

$(\text{PredConOne}\ a \supset (p\ \text{Under}_e\ x,a \vee p\ \text{Under}_e\ x,\ulcorner\text{-}a\urcorner))$.

The "meaning" of a name or Russellian description of x for p at e might then be taken as the full array of PredConOne's under which p takes x at e. Let

'$p\ \text{Des}_e\ b,x$'

express that p correctly designates x by the proper name or Russellian description b at e. Then we may let

'meaning (b,p,e)' abbreviate '$\{a \ni (\text{E}x)\ (p\ \text{Des}_e\ b,x\ .\ p\ \text{Under}_e\ x,a)\}$'.

The meaning of b for p at e is then the virtual class of all one-place predicate-constants under which p takes the designatum of b at e. (Note here the metalinguistic use of a virtual class of expressions as well as the relativity of 'meaning' to both a person and an occasion.) Further reflections are needed to extend this kind of handling of 'meaning' to predicate-constants, including relational ones. Note that although the treatment is pragmatic, there is no swallowing up of meaning into an abyss of subjectivity. Some "meanings" easily generalize to the social group, so that the methods suggested form a basis for extensions into socio-linguistics and the study of intersubjectivity. Note, however, that nothing especially new here emerges about negation.[17]

Already the use of 'occasions' and of special variables for them ('*e*' with or without primes or numerical subscripts) has crept into this discussion. The use of these brings us at once into the heart of *event-logic*, in which the values for the variables are events or occasions, states, acts, processes, and the like. The shortest contexts in which such a variable may occur are those containing an *event-descriptive predicate*. Let '<swim>' be such a predicate so that

'<swim> *e*'

express that *e* is an act of swimming. How long an act temporally? Well, at least long enough for *e* commonly to be reckoned as an act of swimming. The study of event-descriptive predicates turns out to be very helpful for the study of gerundives, of change, of adverbs, of tense, and of related subjects.

The study of adjectives and adverbs is of especial interest in its connection with negation. Many such words tend to be "fuzzy" and allow for much slack. For example,

(i) 'Henry is a kind man'

allows of much leeway, in part depending upon speaker and occasion of utterance or assertion. It is convenient to handle such sentences in *scalar* terms somewhat roughly as follows. Let 'K' be the relation of being less-kind-than as between human persons. The use of 'K' provides immediately for the comparative case of the adjective 'kind', the positive and superlative cases then being handled by definition. Let

'h Mid_e K, 'Per''

express that Henry (or h) is placed midway in the scale for K relative to the predicate 'Per' for human persons on occasion *e*. The assertion of (i), if tense is disregarded, involves in part the speaker's placing Henry somewhere in the middle of the scale for K relative to 'Per'. The scale is intended to be an objective one, depending not just upon your or my whim as to who is kinder than whom, but upon well-agreed upon types of attitudes and behavior. We may distinguish as many places on the scale as we wish, High, Mid, Low, VeryHigh, VeryLow, VeryVeryHigh, VeryVeryLow, and so on. Consider now the negation

'$\sim$ h Mid_e K, 'Per''.

This cannot be read as

'Henry is not a kind man',

for he could be placed VeryHigh in the scale but not in the middle and yet would correctly be said to be kind. A law of excluded middle does obtain, however, in the sense that

$$(\text{h Mid}_e \text{ K,'Per' v} \sim \text{h Mid}_e \text{ K,'Per'})$$

for appropriate *e*. Similar considerations obtain if K is construed as applying to acts rather than persons.

The handling of scalar adverbs is similar, but it is now events, changes, acts, and so on, that are placed on appropriate scales. This handling of adverbs seems much simpler and more direct than those proposed by Reichenbach or Montague.[18]

Metalinguistic considerations have been brought into the construction here or there, but thus far no metametalinguistic ones. At almost any point, however, it may be necessary to turn to a metalanguage of the metalanguage of whatever has been developed up to that point. This should cause no alarm, the hierarchy of languages of ascending level being open-ended.

In this brief sketch of a logico-semiotical system, three kinds of entities have been admitted for the three kinds of variables respectively: individuals, linguistic expressions of the language or metalanguage at any level, and events, states, acts, and so on. Nothing fundamental *de rerum natura* is to be read into this trichotomy, and the three domains of entities may be unified if desired. Some may even be reducible to others in suitable ways. The sketch is merely of a system that seems to work rather well for the study of the logical forms of many kinds of sentences of a natural language. Also there is no neglect of mental phenomena, such being accomodated by means of appropriate event-descriptive predicates.[19] The resulting theory is of extraordinary power of expressiveness, wider even than that of natural language itself. In all of this, as we have seen, the notion of negation '∼' interanimates with the other truth-functional connectives, with the quantifiers, with the non-logical predicates, and with special predicates such as 'P' for part to whole, 'C' for concatenation, and so on, in so fundamental a way that almost nothing of import in the system can be said without it. Any basic distinction between positive and negative statements has pretty much vanished in thin air, as already noted — where by a positive statement one means a wholly positive one, so to speak, containing no inner occurrences of '∼' when expanded into primitive notation.

There are many alternative and “deviant” types of logic on the contemporary philosophical marketplace clamoring for attention.[20] It is often thought that these involve kinds of negation differing radically from the classical one. Let us reflect upon a few of these to help decide whether this is the case, or whether the new types of negation are merely further special cases of the classical one. Let us first consider three-valued logic and intuitionism.

The key problem to ask about three-valued logic is: What is the third value? A “value” intermediary between truth and falsehood, we may be told in answer. This is not much of an answer, however, unless we are told a good deal more. If the third value is taken as “neither known or unknown,” or as something of the sort, the characterization is no longer semantical but pragmatical. Very well, let us then study three-valued logic in its proper pragmatical setting. But note that classical negation reigns there supreme, as we have observed above. The third value is then something like indeterminacy as discussed above — but we can now leave the human person out of the picture, in much the fashion that Husserl did.[21] We then have ‘Tr’ and ‘Fls’ as in semantics, and can let

> ‘Indet *a*’ abbreviate ‘(Sent *a* . $\sim$ (E*e*)(<Kn> *e* . (*e* Of$_{\text{Object}}$ *a* v *e* Of$_{\text{Object}}$ $\ulcorner\sim a\urcorner$))’.

A sentence *a* is thus indeterminate — or has the third value — just where there is no act or state of knowing of either it or its negation. Of$_{\text{Object}}$ here is the prepositional relation of *being-the-object-of*.[22] One can then go on to frame axioms about this third value, incorporating them within a semantics containing ‘Tr’ and ‘Fls’, achieving a kind of three-valued logic with a fixed interpretation.

One can develop mathematical intuitionism, in a somewhat similar pragmatic fashion, by letting ‘<Constr> *e*’ express that *e* is a mathematical “proof” or construction of a kind the intuitionist condones. Let then

> ‘*e* Constr *a*’ abbreviate ‘(Sent *a* . <Constr> *e* . *e* Of$_{\text{Object}}$ *a*)’,

expressing that *e* is an intuitionistically acceptable construction of the sentence *a* — *a* being a sentence here of some portion of classical mathematics. The theory of intuitionism then becomes a theory concerning acts of construction. The classical law of excluded middle holds in full generality, so that

$$(a)(e)(\text{Sent } a \supset (e \text{ Constr } a \vee \sim e \text{ Constr } a)).$$

This principle contrasts of course with one stating that

$$(a)(e)(\text{Sent } a \supset (e \text{ Constr } a \vee e \text{ Constr } \ulcorner \sim a \urcorner)),$$

which of course does *not* hold. Intuitionistic assertion of *a* is then to the effect that there is, or has been, a construction of *a*, and intuitionistic denial is that there is a construction of $\ulcorner \sim a \urcorner$. The task of framing axioms for intuitionism on this basis remains, but is surely possible in principle.

Somewhat similar considerations obtain with respect to *epistemic* logics, concerned with believing, knowing, and the like, with *deontic* logics, concerned with obligation and permission, with the logic governing *perception* and *conception*, with the logic of *ostensive acts*, and with the logic of *assertion, questioning, exclaiming, subjunctivizing*, and *commanding*. *Modal* and *quantum*-logics are of another ilk. The problem of interpreting the former within suitably formulated metalanguages raises its own special difficulties and seems not to have been faced up to sufficiently by modal logicians. Likewise the problem of formulating quantum-logic in full detail with full specification of all the non-logical predicates needed seems not yet to have been accomplished. When these tasks are squarely faced up to, it will very likely turn out to be seen that the metalanguages needed will be of the usual two-valued classical kind, with '~' playing its accustomed role therein.

As goes '~', so goes many another negative word or phrase in natural language: *no, never, nothing, nobody, hardly any, rarely, none, scarcely, seldom*, and so on and on. Some of the logical features governing these words or phrases are well known[23] and need now to be codified and integrated within some over-arching theory.[24] Work in this direction is of course partly empirical, but also depends to some extent on sheer logical analysis in terms of the logical forms of the sentences containing the given word or words.

The unwary reader might be tempted to think that the foregoing sketch of a system is much too formal and remote from the commonalities of philosophical parlance. All this formalism has nothing to do with *my* philosophical reflections or with the philosophical terms *I* use — it might be objected. In reply, it should be pointed out that the usual kinds of philosophical parlance are in sentences of natural language. The foregoing theory — if not just this one, then something like it — is more and more being shown to be able to provide logical forms for all natural sentences. To claim this is not to claim that all the terms a given philosopher uses are thereby subjected to analysis. The problem of the logical status of these terms is the responsibility of the given philosopher. All the philosopher or logician of natural language can do is to provide logical forms for the sentences in which they occur. Each philosopher must supply the rest — the denotata or designata of his or her terms, and the basic meaning-postulates (in Carnap's phrase[25]) governing them. Logically speaking, this is the very first duty of every philosophical writer. Some meaning-postulates for the foregoing system have been suggested here and there, but many more are needed. Much work in supplying such remains ahead.

As a matter of fact, it is an extremely difficult matter, in most areas of systematic thought, even to approximate an adequate listing of meaning-postulates. There are usually boundary conditions, statements of empirical fact, that not only should be taken into account but determine in large part the very character of the inquiry. Here it is only a *partially formalized language-structure* that is relevant.[26] A partially formalized language is like an ordinary logical system, but with empirical statements of fact (taken as hypotheses where needed in deductions) playing a role quite as important as fundamental logical axioms and the non-logical meaning-postulates. The problem of accurately stating the matters of fact remains, however, as acute a one as stating the axioms and postulates. It is here also that the use of logical forms steps in to be of genuine help.

Let us close by extending an important off-hand remark by Frege in his famous paper "Negation"[27] — extending it in such a way as to recognize the philosophical importance of only partially formalized languages. "In any science [or systematic inquiry] in which it is a question of conformity to laws [and accurate statement of boundary conditions], the thing that must always be asked is: What technical expressions are necessary, or at least useful, in order to give precise expression to the laws [and boundary statements] of this science [or discipline]?" We should not only ask this question but seek to answer it as best we may, given the logical and other means available at the time. "What does not stand this test", Frege concludes, "is from evil." To try to pass this test without giving the little '$\sim$' for negation its proper role would be an even greater evil still.

CHAPTER TWO

On logical semiotics and logistic grammar: relations, roles, representations, and rules

> "On ne jette des pierres qu'à l'arbre chargé de fruits."

By a logistic grammar of a natural language let us understand what could also be referred to as the logic or logical semiotics of that language. The very phrase 'the logic of so and so' has a kind of tradition built into it and usually indicates an *applied* logic of the given subject-matter. Further, the study of logic and of the grammar of a natural language have developed more or less hand in hand over the centuries, so that "the logic of natural language" and a well-systematized grammar are almost one and the same. The two in fact have been so intimately associated historically, that it is astonishing they have become so dissociated in recent years. Logic itself, including logical semiotics, has taken tremendous steps forward, and the time seems now ripe to show how these steps are of interest for the grammarian. It might be thought that this has already been done, but, alas, it has not. This is not the occasion to attempt to explore why, which will have to be left for future historians and sociologists of knowledge to study the near-scandal involved.

It is now commonly accepted that "[a]s grammar originally grew out of logic, the former drew from the latter many of its terms," as a wise grammarian of yesteryear has noted.[1] "There are, moreover, historical reasons why a thorough knowledge of grammar can not be obtained without a previous acquaintance of logic. Grammar grew up out of logic, and still retains some of the features of its origin. The early Greek grammarians transferred the terms of logic, many of them, to grammar. These same terms are still employed by some of the best German, French, and English grammarians.

Thus the term *predicate*, used in grammar, was derived from logic. In order, therefore, to understand those terms in their full and exact meaning, the study of logical forms is a prerequisite." And *a forteriori*, of course, if the newer developments in logic itself are taken into consideration.

In several of the author's recent publications, the beginnings of a logico-grammatical theory of wide applicability have been sketched.[2] In his review of *Logico-Linguistic Papers*, the linguist Henry Hiż[3], commenting on these beginnings, remarked that "Martin envisages that every sentence has its logical form. The logical form is in a language of elementary [or first-order] logic augmented by mereology, i.e., a theory of part-whole relations and by such concepts as *event, doing, agent,.patient, give, now, before, during, person,* etc. [italics added] for which one has to accept suitable meanings postulates (as, e.g., that every event which is a giving is a doing). These are not English words *event, do, give* but concepts which are represented in the general logic. The universal logic is conceived as containing syntactical, semantical and pragmatical [— i.e., semiotical —] concepts and as representing all languages. There are transformations from the universal logical forms to actual sentences in the particular natural languages; from the same logical form there are different transformations to the sentences of different languages. It is a logician's version of the deep structure; it is perhaps sounder and certainly of much older origin than the popular version."

Hiż speaks here of the envisagement that "every sentence has its logical form" without calling attention to the enormous number and variety of types of sentences for which logical forms have actually been exhibited, in the volume under review and elsewhere. The use of 'envisages' here seems thus rather ill-chosen. Envisagement takes place only in the absence of achievement. Also Hiż speaks of merology as a "theory of part-whole relations" in the plural, whereas the use of only one suffices for most of the purposes discussed. Others may be introduced for special purposes, e.g., for providing for atomic physics. Note also Hiż's highly ambiguous use of 'represents'. An English word is represented in the universal logic in a very different way from that in which natural languages are represented in it. As a technical term, 'represents' is often used by linguists in still a third sense, that in which a specific "reading" of a natural sentence is represented by a logical form. The first and third senses of 'represent' seem useful and may be retained, the second being somewhat vague and askew. And, of course, meaning-postulates must play their role *überhaupt*, even in the underlying mereology. Meaning-postulates are needed to characterize the part-whole relation adapted (i.e., to lay down its salient properties) as well as to characterize all the other primitives needed.

Hiż accepts mereology as a branch of logic, probably following his mentor Leśniewski in this respect. It could be argued cogently, however, that the part-whole relation on which any mereology is built is not a "logical relation" but rather an empirical or descriptive one. The meaning-postulates lay down very general properties of it, e.g., its total reflexivity and transitivity. For a full theory of any part-whole relation, however, a great deal more is required. If it is to be regarded as involving space or time or space-time or whatever, in some way, a good deal of spelling out of this fact is needed in the theory. If mental events and states are admitted, this too should be spelled out. The resulting system must thus be a good deal more inclusive than what is ordinarily called 'a logic'; it is applied logic of a wide extensiveness. For all of this, further meaning-postulates are needed at every step of development. Hiż seems to speak disparagingly of these, as though they are not needed in grammatical studies. If he is correct in this, it behooves him to show, precisely and in explicit detail, how they may be dispensed with.

Nor is a "logician's version" of deep structure to be lightly tossed aside. It is doubtful that any linguist has ever given an even remotely satisfactory account of what a "linguist's version" of deep structure is supposed to be. The logician's version has at least the minimal distinction of having been spelled out in some detail, at least in a preliminary fashion — as e.g., in the twenty-one desiderata laid down in the last chapter, "On the Very Idea of a Logical Form," of *Semiotics and Linguistic Structure*. To date no one seems to have questioned at any depth the desirability of any of these, nor does anyone seem to have added or subtracted from the list or to have come forward with an alternative one in its place.

If the logician's version is "of much older origin" than the popular alternative, this is so much to the good. The greatness of tradition is not to be disregarded lightly, and we all have much to learn from it in most areas of inquiry. To disregard its merits is often merely to repeat the same old errors. The real secret is, of course, how to distinguish the merits of tradition from its defects and shortcomings.

"The strictly logical part of the universal logic of Martin," Hiż goes on, "is weaker than the set theory and higher-order logic used by Montague[4]; the semantics and the pragmatics are stronger in Martin's theory. There is no point of arguing which choice is better in principle. If any of them succeeds to cover a large area of some natural languages in explaining their semantic nuances, our understanding of the phenomenon of language will be enhanced. If both succeed — so much the better. Martin's actual work is often of more interest to the linguist because Martin, mostly under the influence of Zellig Harris, examines more difficult grammatical problems than Montague did. This is more to the credit of Martin than of his theory, though the theory shows some of its applicability."

Now the words 'stronger' and 'weaker' are notoriously tricky ones as used in comparing logical systems or theories. Surely Hiż's first contention here is correct, on almost any acceptable construal of 'weaker'. The second contention, however, is dubious. Montague's semantics is no doubt the stronger one, but the strictly pragmatical notions needed are pretty much the same. Hiż makes no reference to the detailed discussion of the interrelation of these two approaches to the exact study of language, in "Why I Am Not a Montaguean Grammarian."[5] This is not the occasion to review the niceties involved or to discuss further ones, but merely to record Hiż's neglect of them. It is probably true, on the whole, that the "universal logic" approach can handle "more difficult grammatical problems" than Montague was able to. However, Hiż's reason for stating this are surely obscure — "more to the credit of Martin than of his theory." To give any point to this contention, Hiż should call attention to specific instances wherein he thinks the theory somehow fails and Montague's succeeds. Any such instances would be a welcome challenge to try to do better.

Hiż goes on to note that "of course, Martin's and Montague's grammars are not the only possibilites. While they translate sentences of a natural language into formulae of a logical language in order to perform reasonings, draw consequences according to the laws of logic, and then return to the natural language; logico-linguistics may proceed ... differently." Very well, let it do so. But how? Hiż sketches his own method as follows. "Instead of making a universal logic one can attempt to do the logic of a natural language in the language itself." Yes, but not without encountering the dangers of illicit self-reference presumably lurking therein. "After all, the logic of mathematics was done in the language of mathematics" — a comment vague to the point of being meaningless. If by 'mathematics' one means numerical mathematics — the theory of functions of a complex number, say — and by 'the logic of mathematics' we mean the theory of truth-functions and quantifiers by means of which the language of mathematics may be formalized, then the latter is surely "contained" in the language of mathematics. To contend that "the logic of mathematics was done in the language of mathematics," however, suggests that it was *first* done so. This is of course historically false. To describe just how the logic of mathematics was first "done" would require a careful historical study, and would most surely show that it was not done just "in the language of mathematics." Much more is involved, the logic of mathematics, no matter how construed, containing a logic *simpliciter* of wide applicability as a part. And in any case, no matter how Hiż's contention is disambiguated, it lends no support whatever, as he apparently wishes it to, to a contention that the logic of natural language should be "done" solely in natural language.

"Martin makes a representation of English words in his logic, just as Montague did. This necessitates rendering of *Adam* as *A* and of *book* as *B*. And it necessitates resolution of ambiguities prior to the representation..." Well, ambiguities must be resolved somewhere along the line of course — the sooner, no doubt, the better. If 'Adam' is taken as a proper name of a person, a suitable subscript '*n*' may be added, so that 'A_n' is taken to designate just such and such a person, the person Adam intended in the given discourse. And similarly 'book' may be disambiguated into some of the many uses listed in the *Oxford English Dictionary*.

This resolution of ambiguities, Hiż states, is such that the "English *give* will have many representations: one for *give* in *Adam gives Ruth The Tempest*, another for *give* in *My accent gave me away*. ... And [one may add:] *Give her my best wishes, Father gave him a spanking, The judge gave me 5 years, It's give and take, I give up*, and many more. Each of the uses of *give* here is different and Martin assumes that we can keep them apart. For each *give* some meaning postulates must be stated." But alas, these contentions are not quite correct. Not all of these uses of *give* need be "kept apart". The main *types* of usage are to be distinguished, essentially perhaps as in the lexicon. Some types may well be definable in terms of others. Expressions for some are to be taken as primitives, others then being definable. Ordinarily the meaning-postulates required will characterize primarily the primitive ones.

We now come to the real nitty-gritty of Hiż's preferred method. "An alternative way to logico-linguistics may take all the grammatical constants (for English not only *if-then, not, every*, but also *-s, the, a, re-, be, have, -ly, -tion, he, she, it, -ing*, and all others with which grammars usually deal) of a language in the same way as logic takes the logical constants and supply some consequence rules based on them." In a logical system, of course, *all* the consequence-rules should be supplied, but not *all* the logical constants need be taken as primitive. And so in logico-linguistics also. Not all the grammatical constants need be taken as primitives, some being definable in terms of others. A few such definitions will be given in a moment.

Let us consider Hiż's two examples of consequence-rules. "[1] If all α's are always β's and γ is a α, then γ is β; or, [2] for some specified class of α, β, and γ, if βαγ, then α *is* β and αγ: for instance, if *Old Martha flirts*, then both *Martha is old* and *Martha flirts*." (Note that in the statement of this example there is confusion of use and mention. What is intended is: *Old Martha flirts* has both *Martha is old* and *Martha flirts* as logical consequences.)

[1] is a consequence rule containing — or "about" or characterizing — the grammatical constants *if-then, all, are, always and, is*, and *a*. It is stated *solo* presumbly as a completely general rule for α, β, and γ of a specified kind. But is not this merely another way of saying that 'α', 'β', and 'γ' must "represent" natural-language words or phrases of a specified kind? [1] seems to be similar to a familiar law of quantification-theory, but actually it is very different. We must think of it rather as a principle of grammar. Do all principles of quantification have to be provided for in the course of a full peroration of Hiż's theory? If so, we must be sure this is done. If not, just which ones may be dropped? And why? The second example, [2], is of a similar kind. The consequence-rule here too is not very informative without the explicit specification of the class to which α, β, and γ are supposed to belong.

Of course *Old Martha flirts* is highly ambiguous. A few "readings" are: Old Martha (usually, frequently, habitually) flirts, Old Martha is (now) in process of flirting, OLD Martha flirts (but the other Martha or Marthas do not), Old MARTHA flirts (but old Beatrice and so on do not), Old Martha FLIRTS (but goes no further) — to mention only a few. To capture all of these, Hiż's rule [2] is of course not sufficient but must be supplemented for each reading. His rule thus, at the very simplest, will presumbly have to be a very lengthy disjunction. Also to spell out in full the restrictions on α, β, and γ in order for [2] to hold would be next to impossible. It is difficult enough to disambiguate the parent sentence *ab initio* before seeking a special rule of consequence concerning it.

Hiż's method seems to depend much on the fact that "instead of representing English or Italian words by symbols in logic, e.g., both *give* and *dona* by G, one may apply logic to the language as it is, to English or Italian directly." Is not too much being made here of the notion of "a symbol in logic"? When doing Italian grammar, use 'dona' as a technical expression and when doing English grammar, use 'give'. The use of 'G' instead of 'give' is quite beside the point. As in logic generally, the use of special symbols is merely convenient shorthand.

The chief defect of Hiż's "alternative" in logico-linguistics is that he virtually throws away nearly 2000 years of progress in the development of logic. He wants to start from scratch, taking all grammatical constants on a par with logical ones, and then frame consequence-rules for them *en masse*. To do this, nowhere along the line using the hard currency of logical principles as a helpmeet or guide, is so formidable a task as to be almost hopeless. He does not "divide and conquer," studying simple kinds of linguistic phenomena first and then developing the extensions needed for more complex ones.

Just think for a moment how incredibly complicated and how numerous Hiż's rules of consequence will have to be for even 'not', say, one of the simplest of the "logical" constants. We have seen, in I above, how remote the English 'not' is, in many of its uses, from the simple '$\sim$' of truth-functional logic. Hiż's rules concerning 'not' will have to cover all of the uses considered there, together with all others. In I we have seen how some at least of the various uses of 'not' can be defined in terms of the basic '$\sim$' and hence that the various rules governing these uses are (in part) derivable from the basic logical rules governing '$\sim$'. Hiż does not wish to avail himself of these logical rules, but instead seeks rules of vast generality covering all uses of 'not' simultaneously. This is a grandiose undertaking doomed to failure. Similar remarks obtain for the other so-called logical constants. The real tragedy of Hiż's view is that so little has been done to substantiate it in detail.

Let us go on now to review — in a little more detail than in I above, albeit somewhat synoptically — the essentials of the "universal logic" Hiż rejects in favor of "direct" analysis, in order to see how, in principle, it can provide the wherewithal for logical — including semiotical or metalogical — forms or deep structures for many types of sentences of English. First, we should remind ourselves of the character of the basic logical part of the theory before moving on to syntax, semantics (both designational and intensional), and pragmatics. A grammar of a language is of course in part a metatheory about — or a semiotics of — the language. It, therefore, on a par with other theories and metatheories, contains a basic logic by way of underpinning. The logic should be kept as simple as possible, so as not to intrude in any way upon the resulting theory. If the logic itself is not sufficiently simple — as, e.g., with Montague — it will tend to assert itself excessively; it will dictate features of the theory that the latter neither needs nor should have.

All the basic words of English, in a sense to become clear as we go on, are assumed to be present as non-logical words, as items in the vocabulary of the system. Since English contains its own metalanguages, the syntax, semantics, and pragmatics involved, at whatever level, become parts of English also. A full listing of this entire vocabulary needed will not be attempted here. Various fundamental types of English words, as viewed in terms of the theory, however, will be listed and typical ones within each type picked out. It turns out that these may all be handled as *predicates*, in addition to the connectives, quantifiers, and a few other syncategorematic expressions. The main types of predicates needed are those for the various categories of entities admitted, for some temporal relations, for some structural or "role" relations, for some scalar notions and relations leading to adjectival and adverbial modifiers, for some prepositional relations, for the various verbal "moods," and a few others.

Horne Tooke, in his famous *The Diversions of Purley*, contended apparently that the traditional doctrine of the parts of speech is "absurd and that there are in fact only nouns and verbs."[6] The view here is that there are not even these at the level of deep structure. No, there are nothing but the non-logical predicates, the truth-functional connectives, and some method of handling the quantifiers. To make cogent so radical a contention — in modern tems — is in part the aim of the present discussion.

By a *rule*, then, let us understand what is usually meant in discussions of logic systems, i.e., a meta-statement stipulating that any meaning-postulate (or formula of such and such a shape) is an axiom or theorem, together with meta-statements enumerating rules of inference. The *representation*-problem, in contemporary linguistics, is concerned with assigning or generating logical forms for natural sentences. But each natural word has its sacred *role* to play in the natural sentences in which it can significantly occur. The total number of roles that need be recognized for any given word, and hence for all the words of the language, is finite. Hence there is hope that, in due course of time, a complete enumeration of them could be given. Of course, we are far from such an enumeration at the moment, but it is in the offing. How now are roles to be talked about in the over-all theory? By means of suitable structural or quasi-grammatical *relations*. Some logico-linguists wish to speak of functions here instead. But functions are a special kind of relation. For linguists it is relatively unimportant whether a given relation be functional or not. Insistence on functionality rather than relationality has been the source of much unnecessary obfuscation, and has tended to foist mathematical terminology on linguistics unwittingly. In some such way as this, then, the four items of the sub-title are presumed to be interrelated.

Let us presuppose now the usual first-order logic with virtual classes and relations.[7] Proper names and even Russellian descriptions are to be handled predicatively. Thus 'Soc' may be taken as the one-place predicate 'socratizes' so that

'Soc x'

expresses that x socratizes, i.e., that x is the one and only person who socratizes. A postulate concerning 'Soc' is that

$$\vdash (\mathrm{E}x)(y)(\mathrm{Soc}\ y \equiv y = x).$$

And, similarly, for Russellian descriptions, so that

'$(\iota x\ .\ {-}{-}x{-}{-})\ y$' may be taken as short for '$\{ z \ni (x)({-}{-}x{-}{-} \equiv x = z)\}\ y$'.

('$\{z \ni {-}{-}z{-}{-}\}$' here is the expression for the virtual class of all z's such that --z--.) If in fact it holds that

$$\vdash (\mathrm{E}y)(x)({-}{-}x{-}{-} \equiv x = y),$$

then the definiendum here states that y is the one and only entity x such that --x--.

Another type of one-place predicate of importance for what follows consists of the gerundives. Consider the English word 'walk', and suppose, for the moment, that it is taken as a primitive of the language. Then '<Walk>' is taken as a predicate for being a walk*ing*, thus incorporating the suffix *-ing*. Let

(1) '<Walk> x'

express then that x is a walking, a complete occurrence of the kind we would normally call 'a walking'. And similarly

(2) '<Think> x'

may express that *x* is a complete act of the kind we would call 'a thinking'. And, similarly, for all manner of so-called "verbs." In general, in formulae such as (1) or (2), the *x* is to be regarded as an event, or as belonging to some species of events to be distinguished below.

In terms of gerundives further suffixes and prefixes may be defined. Let us consider the *-s* suffix, where '<Give>' is an event-descriptive predicate for acts of giving. Then

> 'gives' may be taken as short for '{*e* ∋ (<Give> *e* . *e* During *now*)}',

where During is a fundamental temporal relation and 'now' is the deictic expression for the present moment, to provide for some — "third-person singular" — context in which it may occur to express the present tense. This definition may of course be generalized to include all uses of '*-s*' that express the present tense. And, similarly, for all the other suffices and prefixes of Hiż's list. Hiż wants "consequence rules" given based on his primitives. When explicitly supplied, however, these rules will turn out to be either meaning-postulates or rules of consequence derived from them making use of suitable definitions.

No theory of language could get very far without containing a syntax, a syntax of its own object-language part, of the syntax of this syntax, and so on. This syntax can be taken as of the familiar inscriptional type[8], but now writ large for the entirety of the object-language part. For this purpose a full listing of the primitive lexical words of the language is needed. Although no such list is yet available, one ought to be forthcoming before too long.

To form the shape-descriptive predicate for a given English word, let us follow the customary practice of putting that word in single quotes. Thus we may write "Soc" for the predicate 'Soc'. And similarly for the other predicates and the syncategorematica such as '(' and ')' and even the truth-functional 'v', '~', and so on. The semantics may be handled in terms of 'Den' for denotation in the sense that a one-place predicate denotes severally the objects falling under it.[9] The other semantical relations needed are then definable. Intensional semantics may be handled in terms of the relation Under, where

> '$P\ \text{Under}_e\ x,a$'

is taken to express that person *p* consciously, in an intensional act *e*, takes the entity x under the one-place predicate a, i.e., regards a as denoting x (even perhaps where a does not correctly denote x in accord with the rules of the language).[10] Theoretical semantics may in fact be regarded as the full theory characterizing 'Den' and 'Under' taken as the fundamental primitives.

Note that the relation Under is already a pragmatical one, not merely a semantical one in any narrow sense. In fact, intensional semantics, as conceived here, is already a branch of pragmatics. (See III below.) Other important pragmatical predicates are 'Acpt' for acceptance, 'Appr' for apprehends, 'Asrt' for assertion, 'Utt' for utterance, 'Qstn' for questioning, 'Cmnd' for commanding, 'Excl' for exclaiming, 'Sbjnct' for acts of subjunctivizing, and so on. The latter will be recognized as predicates for handling the various moods of assertion, questioning, and so on. Thus 'p Sbjnct_e a' may express that p, in an intensional act e, entertains the declarative sentence a subjunctively. There is no mystique about subjunctives not to be captured by the study of a certain way in which declaratives may be entertained.

The basic predicates needed for the fundamental *categories* of entities are to include the following: 'Ent' for all types of entities in general, 'Obj' for spatio-temporal objects, 'Abstls' for abstractuals or abstract objects in general, 'Ev' for events of all kinds, 'MentEv' for specifically mental events, 'Act' for acts or actions, 'Proc' for processes, 'Change' for changes, 'Per' for persons, 'Pl' for places, 'T' for times, 'Sh' for shapes or sign-designs, and 'Inscr' for inscriptions or sign-events. These categories are not intended to be mutually exclusive or jointly exhaustive, but merely some useful ones to help get us started on the right track.

The basic *temporal* predicate needed is '$\text{Before}_{\text{Time}}$' where '$e_1$ $\text{Before}_{\text{Time}}$ e_2' expresses that the event e_1 occurs wholly before the event e_2 in a suitable temporal ordering. Some definable predicates are then 'During' where 'e_1 During e_2' expresses that e_1 takes place during the time-span of e_2, '$\text{Throughout}_{\text{Time}}$' where '$e_1$ $\text{Throughout}_{\text{Time}}$ e_2' expresses that e_1 takes place throughout the duration of e_2 in the sense that e_1 During e_2 and e_2 During e_1, 'Mom' for being a moment, 'At_{Mom}' where 'e_1 At_{Mom} e_2' expresses that e_1 takes place during the moment e_2, '$\text{Again}_{\text{Time}}$' where '$e_1$ $\text{Again}_{\text{Time}}$ e_2, F' expresses that e_1 and e_2 are both members of a virtual class of events F and e_2 $\text{Before}_{\text{Time}}$ e_1, and so on. And of course a suitable deictic 'now' is needed for the present moment, or, alternatively where 'sp Now e' expresses that the speaker takes e as the present moment.

A leading feature of the grammar to be sketched will be that the *semantical role* of words in many types of sentences will be explicitly indicated, as already noted. Typical roles are those of being a (or the) agent, patient, instrument used, goal, source or origin, location, duration, or result of an action. Ultimately a complete list of all such roles should be provided, all the roles, that is, which the language is able to distinguish by having separate expressions for. Normally these roles will be handled by means of certain relations, usually dyadic ones. For example,

'*p* Agent *e*'

is to express that person *p* is the or an agent of the act or action *e*. Such relations, 'role-relations' we might call them, will be useful in forming answers to questions made with the *wh*-words: *where, when, who, what, why, how*, and so on. Some additional role-predicates are 'Patient' for being the patient or recipient of an action, 'Object' for being the direct object of, or the entity acted upon in, an action, 'Instrument' for being the instrument used in an action, 'Goal' in the sense that 'p Goal e_1, e_2, a' expresses that person p performs e_1 with the goal of having e_2 take place as under the description *a*, 'Location' in the sense that an event takes place at such and such a space-time locale, 'Result' for the relation of being the result of an action or event or state — and so on and on.

Very closely related to the role-predicates are the *prepositional* ones. In English there are about 75 prepositions in all, with many of them used in many different senses. On the whole prepositions are little mono- or bi-syllabic words of an importance in grammar way beyond their length. Many, perhaps most, of them may be defined away, but their grammatical importance is in no way diminished by this fact. A few prepositional predicates may be listed as follows: '$\mathrm{At}_{\mathrm{Position}}$' in the sense that '$e$ $\mathrm{At}_{\mathrm{Position}}$ x' expresses that that *e* takes place at the place *x*, 'e $\mathrm{To}_{\mathrm{AchievedDestination}}$ x' (as in, e.g., 'Tom went to Cambridge'), '$\mathrm{Over}_{\mathrm{Passage}}$' (as in 'He jumped over a ditch'), '$\mathrm{BecauseOf}_{\mathrm{Cause}}$' (as in 'The price of bread is high because of the drought'), and so on and on.[11]

For the handling of what are traditionally called 'adjectives' and 'adverbs', fundamental use is to be made of some *comparative* relations. Let us consider an example of each. First, let us reflect again upon Hiż's use of 'old' in 'Old Martha flirts'. Let

'*p* Younger-Than *q*'

express that person *p* is younger than *q*. Consider now an intensional scale a speaker of 'Old Martha flirts' has in mind when he asserts (or accepts or whatever) this sentence. 'Old' here has a good deal of slack built into it. How old must Martha be to be called 'old Martha'?

Let

'*p* High$_e$ Younger-Than, 'HFLM''

express now that *p* is taken in an intensional act *e* to be high in the scale of certain persons placed or graded with respect to being younger-than, the youngest at the bottom. To be graded High in such a scale is to be taken as old, to be graded low is to be taken as young, to be placed somewhere in the middle is to be taken as neither, to be graded very high is to be taken as very old, very low, very young, and so on. The "certain persons" here are those to whom the predicate 'HFLM' for human females like Martha (as to general health, social position, and so on) applies. 'Old' as used here is thus very different from 'old' as ascribed to elephants, to jets, and so on.

Hiż's sentence in one of its readings may thus be given the form

(3) '(E*e*)(E*e'*) (Martha High$_e$ Younger-Than,'HFLM' . Martha Agent *e'*. <Flirt> *e'* . *e'* During now)'.

From this, as Hiż wants it to, it follows both that Martha is old and that Martha flirts, in accord with the logic of quantification.

Note that 'Old Martha doesn't flirt' cannot be given the negation of the form (3), but rather

(3') '(E*e*) (E*e'*) (Martha High$_e$ Younger-Than, 'HFLM' . Martha Agent *e'* . ~ <Flirt> *e'* . *e'* During now)',

from which it follows that Martha is old and that Martha doesn't flirt. But consider 'It is false that old Martha flirts', where 'false' is taken not in the semantical sense but in the sense of Carnap's absolute concept.[12] Here there is considerable leaway as to what determines the falsity. It may be that Martha is not old, that not she but somewhat else is the agent of flirting or of a non-flirting, or that a flirting does not now take place. If Martha is not old, one would not wish to accept *Old Martha flirts* as true. Consider the following dialogue.

Mr. A. Old Martha flirts
Mr. B. No, Martha is not old. She is only 25.
Mr. A. Oh. I thought she was much older. Anyhow, she flirts.

Hiż too should show us how to provide for such a dialogue, by citing suitable consequence-rules.

Note how pliable the method here is in providing for the various ways in which a sentence may be false. Sentences fail of truth in many ways. For Hiż, presumably, 'Old Martha flirts' may fail of truth only where Martha is not old or if Martha does not flirt or both. The form (3), however, may fail in (2^4-1) ways depending upon whether 3, only 2, or only one of the four conjuncts in (3) fail to hold. The case where all 4 fail would hold trivially.

Adverbials may be handled similarly, but with events or actions or changes or whatever arranged in appropriate scales. Let us consider 'Old Martha flirts subtly'. To handle this we need a definition of the suffix '*-ly*' in the appropriate context. Let

'e_1 Les-Subtle-Than e_2, F'

express that e_1 is less-subtle-than e_2 where both are members of the virtual class F of actions. Here again we need the notion of 'High', so that

'e' High_e Less-Subtle-Than, '$\{e_1 \ni$ <Flirt> $e_1\}$''

expresses that e' is graded high in the scale for being less-subtle-than as confined to the predicate for acts of flirting. This clause we can use as a definiens in a definition of 'subtly (for flirtings)'. That old Martha flirts subtly can then be captured merely by adding this clause as an inner conjunct in (3).

Nothing has been said thus far about mereology or the calculus of individuals as a useful adjunct to the underlying logic. Mereology is the study of part to whole. It is useful to add 'P' for such a relation, if only for purposes of definition. The notion of being a moment, for example, mentioned above, may be regarded as any (non-null) time-slab that is a part of all its (non-null) parts. We thus let 'x P y' express that x is a part of y and add 'P' to the list of logico-linguistic primitives.

The so-called mass-nouns such as 'water', 'oil', 'butter', and so on, may now be introduced. These too may be handled predicatively in two ways. We can let 'W x' express that x is a portion of water and then define

'water' as '$(\iota x \,.\, (y)\,(\text{W}\, y \supset y\, \text{P}\, x))$'.

Or we may let 'Wx' express that x is all of the world's water (past, present, and future), and then express that y is a portion or bit of water by

'$(\text{E}x)\,(y\, \text{P}\, x \,.\, \text{W}\, x)$'.

And similarly — in broad outline — for other mass-terms.

As Zellig Harris and Hiż have repeatedly emphasized, sentences do not in general come singly but in texts or whole discourses. Sentence "interanimate" each other within a text and components of them cross-refer to each other in various ways. A fundamental relation CrRef for cross-reference is needed, e.g., to handle 'he' in the following text.

'Henry entered the room. He was angry.'

Here 'he' is taken to bear CrRef to 'Henry'. Also punctuation marks must be provided for, so that 'Period', 'QstnMrk', 'Colon', 'Semicolon', and 'ExclmPt' are needed as additional syntactial predicates. Some uses of these may be definable, e.g., some uses of a period may be defined away in terms of the truth-functional '.' for 'and'. Some contexts containing '?' may be defined, e.g., where 'Sent *a*' expresses that *a* is an object-language sentence,

'*a*?' may abbreviate '(Sent *a* . (E*p*) (E*e*) (*p* Agent *e* . <Qstn> *e* . *a* Object *e*))',

so that *a* is questioned just where it is a sentence and an object of some act of questioning by someone. '----?' may then be defined by disquotation. The logic of questions may take on a very explicit form when approached in this way. And similarly for the others.

A host of further relations, of a somewhat special sort, are also needed in any logistic grammar. A few such are the following. We may let 'p Able_e e', a' express that e is a state of p's having the ability to do e' as described by a; 'p Oblgd_e e',a,C' that e is a state of obligation for p to do e' as described by a in accord with some code C, moral or legal or whatever, and similarly 'p Prmtd_e e',a,C' for permission; 'e_1 Approp_e e_2' that e is a state of e_1's being appropriate for e_2; 'p Prfrs_e e_1,a,e_2,b' that p in an act or state e prefers e_1 under a over e_2 under b; 'p Ref_e e',a' that p in act e refers to e' by means of a; and so on. Each such predicate requires, of course, very special study and a full theory of each is needed.

A word now about the lexicon. Only a few years ago, the mere suggestion that a lexicon for a natural language should be based on some notions picked out as primitive was scoffed at. Recent work in lexicography has shown, however, that a lexicon should be essentially like that of a logical system — a non-circular list of definitions, or conventions of notational abbrevation,in terms ultimately of the primitives chosen. Of course, there is more to a lexicon than this. Important progress in lexicography is being made, however, and the subject promises to become an integral part of logico-linguistics in due time.

Hiż's comment concerning the transformation rules is essentially correct. "There are transformations from the universal logical forms to actual sentences in the particular natural languages [and back again]; from the same logical form there are different transformations to the sentences of different languages." For the most part, transformations become merely further extensions of the underlying logical rules, making use of appropriate definitions. If only one natural language, English, is considered, as above, the English words are themselves words of the universal grammar. If more than one natural language is considered, the natural-language words must be "represented" by appropriate special symbols, as Hiż correctly observes.

It might appear to some readers that at some point in the development the underlying first-order logic will have to be strengthened. A second-order logic may be needed, or even a full theory of types. If so, the logic may be taken as the ordinal mereology of a previous paper, which still recognizes only individuals as values for variables but with the admission of *ordinal* individuals among them.[13] Ordinal mereology can be given sufficient power to incorporate a full theory of types if needed. Thus no fear need arise concerning the requirement that the underlying logic, with individuals and individuals only as the value for the variables, should be of first order.

A further word of explanation is needed concerning the intensional notion 'Under', which may be interpreted in many ways, two of which are of especial importance. It may be taken normatively, so that 'p $\mathrm{Under}_e\, x,a$' expresses that, in the intensional act e, person p takes x under the description a normatively, or in accord with the semantical rules of the system. Perhaps it would be better to express this by

'p $\mathrm{NorUnder}_e\, x,a$',

where the added 'Nor' is to suggest 'normative'. The other interpretation for 'Under' allows for greater play of usage on p's part, so to speak, reflecting not normative usage only but also p's particular idiolect and noetic or even emotive state. For this we could use

'p $\mathrm{ActlUnder}_e\, x,a$',

to express that p, in e, actually takes x under a in accord with his or own whims, as it were. The restraints on this construal for 'Under' are of course much less severe than upon 'NorUnder', the former reflecting actual usage with its many deviations from the norms of the language.

For example, for 'NorUnder' we have a Consistency Principle, that

$$\vdash (p)(e)(x)(a)(p\ \mathrm{NorUnder}_e\, x,a \supset \sim p\ \mathrm{NorUnder}_e\, x,\ulcorner\text{-}a\urcorner),$$

where ⌜-*a*⌝ is the Boolean negative of the predicate *a*. Such a principle would be inappropriate for 'ActUnder'. Of course if it should obtain that

$$(p \text{ ActulUnder}_e\, x,a\ .\ p \text{ ActlUnder}_e\, x,\ulcorner\text{-}a\urcorner)$$

for given *p*, *e*, *x*, and *a*, and the inconsistency were pointed out to *p*, he or she might, immediately or after due deliberation, wish to make amends. If not, *p*'s linguistic behavior could not be regarded as being in accord with normative standards and with the correct use of '-'.

Rather similar considerations obtain for other pragmatical predicates, such as 'Acpt' for acceptance, 'Ref' for reference, and 'Prfrs' for prefers. The distinction between the normative and the actual has perhaps not always been sharply enough drawn in some of the author's previous writings. Strictly, however, it is desirable to do so in order to differentiate between normative intensionality (with an 's') and actual intentionality (with a 't').

The intent of the preceding discussion has been merely to pull together into a unified framework a multitude of predicates needed in any attempt to provide a basis for a grammar of natural language on logical principles. Hiż himself, somewhere along the line, will have to face up to providing for, and handling, all of these predicates in some fashion or other. Of course much work remains concerning each predicate, the meaning-postulates needed, and so on. In carrying out details, the help of the professional linguist will be needed. Enough has been shown, however, it is hoped, to see how, with further extensions and emendations, the notion of a logistic grammar is no longer a mere dream or vision, but a reality within actual reach with enough cooperative hard work. "Let many [alternative] theories succeed!" Hiż concludes, "We will have gotten that much more insight." But along the way we must try valiantly to keep on the right track, with a certain "logical simplicity" as our guide, lest we fall into the Montaguean morass and the Hiżian complexity.

On every item in the foregoing much detailed analysis is needed. Many philosophers and linguists over the years have been working on such analyses in one way or another. The work of Peirce, Reichenbach, Uriel Weinreich, Donald Davidson, George Lakoff, Raoul Smith, Randolph Quirk and associates, Helen Cartwright, William Cooper, James McCawley, and, of course, Zellig Harris and Hiż himself, *inter alia*, has been especially helpful in drawing up this list. The list is merely tentative and approximative, but even so, appears to be the most extensive of its kind yet attempted.

Some readers may fear that a logistic grammar developed in the manner of the foregoing, "might," in the words of George Hemphill in his *A Mathematical Grammar of English*[14], "reduce a natural linguist disastrously to formulae. What kind of a monster would want to do that to the language of Chaucer, Shakespeare, Milton, the Book of Common Prayer of 1549, the Bible of 1611? The fear in understandable," he notes, "but I fear groundless. As with the Ransomian description of a poem as a structure AND a texture, so also with natural language generally. Grammar is concerned almost wholly with structure; the texture is left almost untouched and can flourish undisturbed. The present writer came to linguistic studies slowly and *via* literary studies, and wishes to assure the reader that mathematical grammar can only illuminate, not blight, the laurels of our true poets." The only difficulty with this admirable statement is that the boundary between structure and "texture" is not sharp, and that texture itself has a structure we should try to understand and characterize. Texture and context, both linguistic and non-linguistic, are yielding to logical analysis quite as much as structure is — but only, of course, where the logical notions used are, as here, of an extent and subtlety far beyond those of logic conceived in the traditional, somewhat narrow sense.

CHAPTER THREE

On Carnap and the origins of systematic pragmatics

"Opera illius mea sunt."

In his seminal monograph *Foundations of Logic and Mathematics* of 1939[1], Carnap notes (pp. 3-4) that "three components have to be distinguished in a situation where language is used...: (1) the action, state, and environment [including "emotional, volitional, and other factors"] of a man who speaks or hears, say, the German word 'blau'; (2) the word 'blau' as an element of the German language (meant here as a specified acoustic (or visual) design which is the common property of the many sounds produced at different time, which may be called the tokens of that design); (3) a certain property of things, viz., the color blue, to which this man — and German-speaking people in general — intends to refer (one usually says, "The man means the color by the word", or "The word means the color for these people", or "... within this language")."

Carnap goes on to note (p. 4) that "the complete theory of language has to study all these three components. We shall call *pragmatics* the field of all those investigations which take into consideration the first component, whether it be alone or in combination with the other components. Other inquiries are made in abstraction from the speaker [and hearer] and deal only with the expressions of the language and their relation to their designata. The field of these studies is called *semantics*. Finally, one may abstract even from the designata and restrict the investigation to formal properties ... of the expressions and relations among them. This field is called *logical syntax*. ..."

In these splendid paragraphs we encounter a characterization of the modern trivium of syntax, semantics, and pragmatics arising from the sea almost full blown like Venus herself. It is interesting to note that this characterization concerns natural languages as the languages under study.

Immediately, however, Carnap goes on to consider (p. 5) a "fictious language B, very poor and simple in its structure, in order to get simple systems of semantical and syntactical rules." B is taken as the *object-language*, and "the language in which the results of the investigation are formulated [is] the *metalanguage*." Only later does Carnap allow the object-language to be a logical system rather than a natural language, but he nowhere reflects in any detail upon the need or possibility of formalizing the metalanguage itself.

In Carnap's trivium, it is also interesting to note, pragmatics comes first in order of comprehensiveness and the other two, semantics and syntax, follow by "abstraction," i.e., by omitting consideration of certain features of language or of language use. From a systematic point of view, however, the order of priority would seem to be the other way around. If 'systematic' here is taken to indicate procedures of language-construction — of formulating language-systems rather than analyzing a natural language in some sense already given — it would seem more natural to start with the simplest area of inquiry and work on to the more complex.

Carnap suggests some pragmatical rules concerning the language B. These are of the form of empirical generalizations, made in a situation in which (p. 5) "we find a group of people speaking [or writing] a language B which we do not understand; nor do they understand ours. After some observation, we discover which words the people use, in which forms of sentences they use them, what these words and sentences are about, on what occasion they are used, what activities are connected with them, etc." We then formulate pragmatical rules to record the results of these observations. We should note, however, how extremely complex such a procedure is. Only a thoroughly-trained field-worker in such study could hope to state any such generalizations with any accuracy. To trace out the path from the spoken phonemes to words to sentences, and from words to things, and from sentences to what they are "about," and from words and sentences to the typical occasions of utterance, and from such occasions to emotional and volitional factors accompanying such use — and so on — requires highly-developed techniques of study, which no amount of arm-chair "common-sense" or "regimentation" can take the place of. The pragmatics of a natural language, as Carnap conceives it, thus embraces the full gambit of what is best done nowadays by socio-linguists, psycho-linguists, and linguistic anthropologists.

The moment he speaks of semantical rules, rather than pragmatical ones, in any detail, Carnap thinks of his language B as being transformed into a language-system B-S. "Strictly speaking," he writes (p. 7), "the rules which we shall lay down, are not rules of the factually given language B; they rather constitute a language-system corresponding to B which we will call the *semantical system B-S*. The language B belongs to the world of facts; it has many properties, some of which we have found, while others are unknown to us. The language B-S, on the other hand, is something constructed by us; it has all and only those properties which we establish by the rules. Nevertheless, we construct B-S not arbitrarily but with regard to the facts about B. Then we may make the empirical statement that the language B is to a certain degree in accordance with the system B-S."

This is an admirable statement of how "facts" are incorporated and expressed within language-systems. It echoes, with a certain difference of emphasis, the famous last paragraph of §2 of *The Logical Syntax of Language*[2], concerning the use in general of logical techniques in the analysis of natural language. This passage is so important for methodology that it should be quoted in full. Carnap states that logical methods and the use of symbolic languages "will not only prove useful in the logical analysis of scientific theories — it will also help in the *logical analysis of the word-languages*... The direct analysis of these [the incredibly complicated word-languages], which has been prevalent hitherto, must inevitably fail, just as a physicist would be frustrated were he from the outset to attempt to relate his laws to natural things — trees, stones, and so on. In the first place, the physicist relates his laws to the simplest of constructed forms; to a thin straight lever, to a simple pendulum, to punctiform masses, etc. Then with the help of laws relating to these constructed forms, he is later in a position to analyze into suitable elements the complicated behavior of real bodies, and thus to control them. One more comparison: the complicated configurations of mountain chains, rivers, frontiers, and the like are most easily represented by the help of geographical coordinates — in other words, by constructed lines not given in nature. In the same way, the syntactical property of a particular word-language, such as English, or of particular classes of word-languages, of a particular sub-language of a word-language, is best represented and investigated with a constructed language which serves as a system of reference."

In his valuable but much-neglected *Introduction to Semantics* of 1942[3], Carnap reiterates the characterization of pragmatics of the earlier monograph, but in a somewhat more precise and developed form. He calls attention to three chief factors in the study of language: the speaker, the expression uttered, and the designatum of the expression. The trivium of pragmatics, semantics, and syntax is explicitly referred to as *semiotics*, and semiotics is divided into pure and applied branches. Here Carnap is mainly concerned with the difference between pure and descriptive semantics. This latter (p. 11) "is the empirical investigation of the semantical features of historically given languages," whereas the former is "the analysis of semantical systems, i.e. systems of semantical rules. Syntax is divided analogously." No such distinction, however, it is interesting to observe, is drawn in pragmatics.

As examples of pragmatical investigations, Carnap mentions the following (p. 10): "a physiological analysis of the processes in the speaking organs and in the nervous system connected with speaking activities; a psychological analysis of the relations between speaking behavior and other behavior; a psychological study of the different connotations of one and the same word for different individuals; ethnological and sociological studies of the speaking habits and their differences in different tribes, different age groups, social strata; a study of the procedures applied by scientists in recording the results of experiments, etc." Observe that all such studies belong to *descriptive* pragmatics and fall within one or another of the sciences mentioned a few paragraphs back, psycho-linguistics, socio-linguistics, and so on.

The concern of *Introduction to Semantics* is of course with semantics, and there are very few references to pragmatics in it. In a later paper, "On Some Concepts of Pragmatics," of 1955[4], Carnap turns his whole attention to pragmatics for the first time. This paper is in part a reply to Chisholm's critique of "Meaning and Synonymy in Natural Languages," also of 1955. In this latter paper Carnap reiterates that pragmatics is "the empirical investigation of historically given *natural languages*." He then immediately speaks of the theory of intensions — as developed in the main body of *Meaning and Necessity* (1947 and 1956) and foreshadowed in *Introduction to Semantics* — as dealing "with concepts like intension [as contrasted with extension], synonymy, analyticity, and related ones." These concepts are here regarded as belonging to pragmatics, but presuppose for their characterization a special branch of *intensional* semantics.

As an example, Carnap considers a certain German-speaking person Karl and the predicates 'blau' and 'Hund'. "It is generally agreed," Carnap writes, "that, on the basis of spontaneous or elicited utterances of a person, the linguist can ascertain whether or not the person is willing to apply a given predicate to a given thing, in other words, whether the predicate denotes the given thing for the given person. By collecting results of this kind, the linguist can determine first, the extension of the predicate 'Hund' within a given region for Karl, that is, the class of the things to which Karl is willing to apply the predicate, second, the extension of the contradictory, that is, the class of those things for which Karl denies the application of 'Hund', and, third, the intermediate class of those things for which Karl is not willing either to affirm or to deny the predicate. The size of the third class indicates the degree of vagueness of the predicate 'Hund', if we disregard for simplicity the effect of Karl's ignorance of relevant facts...."

The question that now rises is (p. 236): "Granted that the linguist can determine the extension of [or class determined by] a given predicate, how can he go about this and determine also its intension?" Carnap defends what he calls *the intensionalist thesis*, that it is possible for the linguist to assign to one and the same predicate different intensions. "For [p. 237] there are more than one and possibly infinitely many properties whose extension within the given region is just the extension determined for the predicate." Note here the key use of 'property', a term of intensional semantics, as contrasted with the extensionalist word 'class'. The intensionalist thesis is thus not to be understood independently of a prior intensionalist semantics, as we have already noted. To the linguist who has a hankering for the clarity of extensionalism and who does not think that the notion of *property* has yet been suitably characterized, the intensionalist thesis may not be convincing. The appeal to pragmatics, although welcome, does not supplant what he might regard as illicit reference to what he regards as an unanalyzed notion. (More will be said on this below.)

Given the prior notion of a semantical intension, no great difficulties have to be faced to relativize it to a given speaker of a given language at a given time. Thus Carnap can readily offer the following definition:

> "Two expressions are *synonymous* in the language *L* for *X* at time *t* if they have the same intension in *L* for *X* at *t*."

The real problem for pragmatics, it might be thought, is to achieve a suitable definition for 'synonymous' altogether independent of the semantical notion of intension.

Let us now turn to Carnap's reflections concerning a more *systematic* approach to pragmatics. In "On Some Concepts of Pragmatics," Carnap comments that "it seems that a more thorough analysis of intension, belief, and related concepts would require a conceptual framework of theoretical pragmatics." He goes on to mention "a few concepts that might come into consideration as a basis of such a framework," symbolizing them explicitly. He lets

(1) 'B(*X*,*t*,*p*)'

express "that the person *X* at the time *t* believes [the "proposition"] that *p*," understood "in a weak sense, as not implying either that *X* is aware of the belief or that he is able to verbalize it." Also he lets

(2) 'T(*X*,*t*,*S*,*L*)'

express "that *X* at *t* takes the sentence *S* of the language *L* to be true (consciously or not)." Also

(3) 'Int (*p*,*S*,*L*,*X*,*t*)'

is to express "that the proposition *p* is the intension of the sentence *S* in the language *L* for *X* at *t*." Two additional symbolizations are introduced, for *assertion* and *utterance*. Let

(4) 'A(*X*,*t*,*S*,*L*)'

express that "*X* at *t* wills deliberately to utter a token of *S* as a sentence of the language *L* in the sense of an assertion." And

(5) 'U(*X*,*t*,*R*)'

means "that *X* at *t* produces with his speaking organs a series of audible sounds *R*," and, where "*R* is a token of *S*,"

(5′) 'U(*X*,*t*,*S*)'

expresses that *X* utters *S* at *t*.

The introduction of the forms (1)-(5′) is already a step toward a systematiziation of pragmatics. Several features of these forms should be noted. All of them contain a parameter for a time. (1) and (3) contain a parameter for an intensional object such as a "proposition." (2), (3), (4), and (5′) contain a parameter for sentences of the language. (2), (3), and (4) contain a parameter for the language, regarded presumably as a value for a variable. (4) and (5′) are read in such a way as to require a "token" of a sentence as well as a sentence regarded as a shape or sign-design. And, finally, in (5), "series of audible sounds" must be taken as values for variables. Any systematization of a pragmatics incorporating (1)-(5′) as sentential forms must thus have a very considerable internal complexity. It must utilize a syntax of both shapes and tokens or inscriptions, as well as semantics containing intensions and no doubt extensions as well. This semantics in turn must be based upon a syntax dealing at least in part with inscriptions.

By 1955 the beginnings of an inscriptional syntax were already available. The Goodman-Quine "Steps toward a Constructive Nominalism"[5] had already appeared in 1947 and the Martin-Woodger "Toward an Inscriptional Semantics"[6] had appeared in 1951. The basis for a systematization of some features of pragmatics was thus available to Carnap at the time. His main interest, however, had shifted to probability and induction, upon which he concentrated for the rest of his life. In 1958 inscriptional semantics was given a deeper and more rigorous formulation, in *Truth and Denotation*[7], in which, however, there was no concern with semantical intensions or Fregean *Sinne*. In *Meaning and Necessity* (1st ed., 1947) Carnap had discussed these latter rather extensively, but without aiming at a rigorous systematization of the metalanguages needed. The discussion there is for the most part semantical, with little referene to pragmatical matters. The question thus remained as to the extent to which these latter can be made to play the role of the former. If a systematic pragmatics is available, can all reference to semantical intensions or *Sinne* in the underlying semantics be dropped without loss?

For Carnap, pragmatics is concerned *only* with natural languages and not with constructed language-systems, as we have already noted. Perhaps a natural language *is* merely a language-system of a very great complexity. For the present, this is a thesis that must be left aside, resting as it does on the subsequent development of logico-linguistics.[8] Carnap's object-language, in *Foundations of Logic and Mathematics*, was already partially systematized or formalized, as we have noted. The natural next step is a systematization of the metalanguage. Carnap's very last paragraph of "On Some Concepts of Pragmatics" was thus a timely one: "There is an urgent need for a system of theoretical pragmatics, not only for psychology and linguistics, but also for analytical philosophy. Since pure semantics is sufficiently developed, the time seems ripe for attempts at constructing tentative outlines of pragmatical systems. Such an outline may first be restricted to small groups of concepts (e.g., those of belief, assertion, and utterance); it may then be developed to include all those concepts needed for discussions in the theory of knowledge and the methodology of science."

In *Toward a Systematic Pragmatics* of 1959[9], the attempt was made to formulate several systems of theoretical pragmatics of the kind Carnap envisaged. The general conception of the nature of pragmatics was in close agreement with his but with some important differences. All the object- and metalanguages considered were systematized, at least partially. No presumption of an underlying intensional semantics was made. In fact, one kind of pragmatics that was discussed presupposed only a syntax, and no semantics whatsoever. When a semantics was presupposed, it was presumed formulated in terms of a special semantical primitive, e.g., 'Des' for a Frege-like relation of designation. The point of view was thus wholly extensional, in a rather thorough-going sense, at both the object and metalinguistic levels.

Let us review now very briefly the main content of *Toward a Systematic Pragmatics*. Chapter I was merely prefatory. In chapter II a Tarski-like formulation of the simple theory of types was given. Let us refer to it as the system T. The syntax developed for T was based on shapes, as in Tarski's *Der Wahrheitsbegriff*[10], but developed by means of Quine's method of *framed ingredients*.[11] The result was a first-order handling of the syntax. The semantics was developed in terms of a primitive 'Des' for designation. The *Rules of Syntax* (pp. 22-23) given were essentially those of Tarski as adapted to the method of framed ingredients. There were three *Rules of Designation* (p. 30), a *Specificity Rule* stipulating that certain constants and abstracts designate certain type-theoretic objects, a *Restrictive Rule* stipulating that if an expression designates an object of type n, it is then a constant of type n, and the *Principle of Uniques* or *Univocality*, that expressions designate at most one object. An especially simple definition of the semantical notion of *truth* was put forward (p. 31) in terms of vacuous abstracts. A sentence a of the object-language is true just where there is a vacuous class-abstract $\ulcorner \hat{x}^1 a \urcorner$ that designates the universal class of type 1, 'x^1' being a variable of type 1. This notion may easily be shown *adequate* in Tarski's sense and the usual principles concerning truth are immediately forthcoming.

Incidentally, it is sometimes objected that taking 'Des' as a semantical primitive is a result of theft over honest toil. The Tarskian method in semantics, it will be recalled, is to *define* all semantical notions within the semantical metalanguage in terms of logical notions "essentially richer" than those of the object-language. These logical notions then play the role, in the Tarskian method, of 'Des' (or 'Den' for denotation, or whatever) in the kind of semantics used here. In either method, it should be noticed, there is "theft" in the use of new primitives. Upholders of the Tarskian method, by parity of their method and for consistency of requirement, should demand also that all *syntactical* notions, as well as all *pragmatical* ones, be definable in terms of "logical" notions. However, this they do not do, being quite happy with new primitives for concatenation, for example, and presumably also for such pragmatical notions as they consider (e.g. Leśniewski's '⊢' for the *assertum*).

In Chapter III a primitive 'Acpt' for acceptance was introduced on the basis of the preceding framework. Acceptance here was construed in essentially the sense of Carnap's 'T' of (2) above, in the sense of "taking to be true." Thus

(6) '*X* Acpt *a*,*t*'

was taken to express the person *X* accepts or takes at time *t* the expression *a* (of the given object-language) to be true (consciously or not). Because of the presence of the time-parameter, an elementary theory of time-flow was presupposed (p. 37). Let us reflect now upon the problem of axiomatizing a pragmatics for Acpt as thus conceived.

Note that (6) was taken to express *actual* acceptance on the part of person *X*, not normative or paradigmatic or "rational" acceptance in some sense. To establish a pragmatical sentence of the form (6) is thus to make use of suitable test-procedures of the kind employed in the empirical sciences. The help of psycho- or socio-linguists would presumably be needed in order to carry out such tests in practice and even to characterize them in general — as Carnap noted. The results of such tests would be framed in some of the *Rules of Acceptance* (pp. 37ff.), more particularly, in *Specificity Rules* stipulating precisely what sentences a given person accepts at given times. The Specifity Rules are thus merely observation-statements stating boundary conditions. A *Restrictive Rule* for acceptance states that any expression accepted is a sentence of the language. A *Continuity Principle* states that *X* accepts *a* at *t* if and only if *X* accepts *a* at every momentary part of *t*.

One might be tempted to think that some *Rationality Rules* should be available concerning negation, disjunction, conjunction, and the like. Thus one might expect a principle that if *X* Acpt *a*,*t* then it is not the case that *X* Acpt *b*,*t*, where *b* is $\ulcorner\sim a\urcorner$, the negation of *a*. Such a principle, however, would characterize *normative* rather than *actual* linguistic behavior. It seemed better, therefore, in *Toward a Systematic Pragmatics*, to consider some class (or virtual class) *F* of sentences of the language, to subject them to the testing proedures, and then to determine whether or not such a normative condition holds.

Let *F* then be some class of sentences containing at least one sentence *a* and its negation, at least sentences *a* and *b* and their disjunction, and also, say, at least *a* and *b* and also ⌜(∼*a* v *b*)⌝ or (by definition) ⌜(*a* ⊃ *b*)⌝. We can then define such notions as being a *normal acceptance pattern* for *X* at *t* for *tilde* ('∼'), for *vee* ('v'), and so on.

In particular we can let

> '*X* AcptNor *tilde,F,t*' abbreviate '(*a*)((*F a* . *F* (*tilde a*)) ⊃ (*X* Acpt *a,t* ⊃ ∼ *X* Acpt (*tilde a*),*t*))'.

The definiendum here may be read: person *X* accepts *tilde* normally (or correctly) at *t* relative to *F*. We could also read this as: the sentences of *F* exhibit a *normal acceptance pattern* for *X* at *t* so far as regards *tilde* (or negation). And similarly for *vee*, where

> '*X* AcptNor *vee,F,t*' abbreviates '(*a*)(*b*)((*F a* . *F b* . *F* (*a vee b*)) ⊃ ((*X* Acpt *a,t* v *X* Acpt *b,t*) ⊃ *X* Acpt (*a vee b*),*t*)).

And similar for many further types of normal acceptance patterns.

The two definitions just given are weaker then those on pp. 40 ff. of *TSP* and provide for alternative notions. The definitions of *TSP* are like the foregoing but have '≡' in place of the right-most occurences of '⊃' in the definientia. The definitions of *TSP* require that *X*'s acceptances be bivalent, like truth itself. The definitions just given allow that *X* need not accept separately either *a*, say, or its negative ⌜∼ *a*⌝. These latter definitions seem better to accord with actual usage in placing a less demanding requirement on "normalcy".

It has already been suggested that in a systematic pragmatics we should be able to dispense with presupposing a theory of semantical intensions. What, after all, is an intension? Clearly, it would seem, an intension should be relativized to a person and a time and reflect in full the person's acceptances of sentences of a certain kind (pp. 40ff.). First, let us consider intensions of an individual constant and all the sentences in a given class or virtual class *F* containing that constant accepted by a given person at a give time. All such sentences would reflect the person's total knowledge of the entity designated by the constant. The intension of the constant — in the sense of its "meaning" — would not extend to the person's total knowledge, but would be determined only by the underlying logic governing that constant. This logic would include the full logic of truth-functions and quantifiers. If the constant is a primitive, the logic would also include all meaning-postulates governing it, in fact, all of the meaning-postulates of the object-language. If the constant is defined, the logic would include its abbreviatory definitions together with the meaning-postulates. Thus the *subjective intension* of a given constant *a* for person *X* relative to a class of sentences *F* at time *t* can be identified with the class of all classes α such that *X* accepts a logical theorem (LogThm) of *F* to the effect that the designatum of that constant is a member of a member of α.

In a similar way, subjective intensions for constants of higher logical type may be introduced (pp. 46f.). And also *intersubjective* intensions, relative to a time and a class of sentences, as a kind of logical product (p. 48) of subjective intensions. And even *objective* intensions (p. 49), relative only to the class of sentences under investigation, may be introduced. And likewise, notions of subjective, intersubjective, and objective *co-intensiveness* may be introduced (pp. 50ff.) supplanting semantical notions of synonymy or sameness of meaning — supplanting them in the sense that once such pragmatical notions are available the semantical notions are no longer needed.

In *TSP* all of these kinds of intensions were called 'quasi-intensions' in order to distinguish them from Carnap's type of semantical intensions. The use of the prefix 'quasi' has been dropped in the foregoing, the various intensions approximate so closely semantical ones. Especially to be observed is the strong sense of 'logical theorem' used in the definientia of the definitions of them. A rather broader kind of purely pragmatical intension arises if the clauses referring to logical theoremhood are dropped. Such intensions would reflect more closely actual linguistic behavior than do those depending on the logic and meaning-postulates of the language. They are thus of greater utility in the study of the logic of natural sentences, as we shall see in a moment.

Of utmost importance here (p. 49) "is the requirement that intensional entities, whatever they may turn out to be, must have *identifiable* parts or members or components in some sense or other. One must be able in concrete cases to specify explicitly and clearly what the intension ... of a term is.... Given any constant *a* of T of given type, the experimenter can enumerate the members of the subjective ... intension of *a* relative to *X*, *F*, and *t*.... And this can be done in terms understandable to anyone who speaks or uses the language T, a clear-cut language-system of the kind commonly presupposed in the sciences. The ... intension can thus be fully and explicitly specified [experimentally] in clear-cut terms in any concrete case. Surely nothing less than this should be required of any logical theory of intension." As a matter of historical fact, the theory suggested seems to have been the first to meet this requirement. And, moreover, the requirement has been observed more in the breach than in the keeping in most subsequent theories.

What, now, about such intensional objects as *propositions*, needed fundamentally by Carnap but avoided altogether in the foregoing? "Most theories of intension ... [that] admit propositions as intensions of sentences [p. 53], admit them as a new kind of entity over and above the intensions of terms.... But this seems to be admitting too much, for such intensions as propositions are intimately related with the intensions of terms and should be definable in terms of them. Consider an atomic sentence of T, say

$$`a_1^1 \varepsilon a_1^2'$$

[expressing that the individual a_1^1 is a member of the class a_1^2]. The intension of this sentence should depend upon the intensions of 'a_1^1' and 'a_1^2', and in fact should be uniquely determined by them. And similarly for the intensions of more complicated sentences as well."

By an atomic proposition corresponding to an atomic sentence of T, then, let us take the *ordered pair* of the intensions of its terms. There will be subjective atomic propositions, intersubjective ones, and objective ones, and further ones of mixed *genre* — nine kinds in all. By a disjunctive proposition corresponding to a disjunction of atomic sentences, we can take a suitable Boolean sum of the atomic propositions corresponding to the disjuncts. By the negative of T, let us take the ordered pair of the intension of the first term with the intension of the negation of the second. And so on. These various notions may be extended to sentences containing quantifiers by various technical devices. By such means the hoary old notion of proposition gives way to a multiplicity of respectable and clear-cut notions for the first time, pragmatical ones now of course, and suitably relativized to the system T.

Many of the foregoing explanations are somewhat loose. They may all be made precise, however, within the pragmatical metalanguage for T, PM^T, the full character of which may be summarized in the following informal stipulation concerning terms and formulae. The object-language is T with a type-theoretic 'ε' as primitive; its syntax is handled by a primitive '⌒'

for concatenation together with shape-descriptive names of the symbols of T; the semantics is based on 'Des' for designation; and 'Acpt' is the sole pragmatical primitive (for the moment) together with 'B' for a temporal before-than relation as between time-slabs. Then we have the following simultaneous recursion.

1. Any variable or constant or abstract of T is a translational term of T.
2. If $\boldsymbol{a}$ is an expressional variable or constant, it is an expressional term of T.
3. If $\boldsymbol{t}$ is a variable or constant for a time-slab, it is a term of T.
4. If $\boldsymbol{x}^n_m$ and $\boldsymbol{y}^{n+1}_k$ are respectively translational terms of type n and $n + 1$, then $\ulcorner \boldsymbol{x}^n_m \varepsilon\, \boldsymbol{x}^{n+1}_k \urcorner$ is a formula.
5. If $\boldsymbol{x}^n_m$ and $\boldsymbol{y}^n_k$ are translational terms of type n, then $\ulcorner \boldsymbol{x}^n_m = \boldsymbol{y}^n_k \urcorner$ is a formula.
6. If $\boldsymbol{a}$ and $\boldsymbol{b}$ are expressional variables, $\ulcorner \boldsymbol{a} = \boldsymbol{b} \urcorner$ is a formula and $\ulcorner(\boldsymbol{a}^\frown \boldsymbol{b})\urcorner$ an expressional term.
7. If $\boldsymbol{t}$ and $\boldsymbol{t}'$ are terms for time-slabs, $\ulcorner \boldsymbol{t}\ \mathrm{B}\ \boldsymbol{t}' \urcorner$ and $\ulcorner \boldsymbol{t} = \boldsymbol{t}' \urcorner$ are formulae.
8. If $\boldsymbol{a}$ is an expressional term and $\boldsymbol{x}^n_m$ a translational one, $\ulcorner \boldsymbol{a}\ \mathrm{Des}\ \boldsymbol{x}^n_m \urcorner$ is a formula.
9. If A is a formula and $\boldsymbol{x}^n_m$ a translational term, $\ulcorner \hat{\boldsymbol{x}}^n_m\, A \urcorner$ is a translational term.
10. If A is a formula, so is $\ulcorner \sim A \urcorner$.
11. If A and B are formulae, so is $\ulcorner(A \vee B)\urcorner$.
12. If α is a variable and A a formula, $\ulcorner(\alpha)A\urcorner$ is a formula.

This informal recursion summarizes, unites, and extends those on p. 17, p. 30, and pp. 34-35 of *TSP*.

This metalanguage PM^{T} is of course a classical one, with expressions construed as shapes. An *inscriptional pragmatics* (p. 65) would construe expressions as inscriptions. Also there can be pragmatical metalanguages of a mixed kind, having both shapes and inscriptions as values for two sorts of syntactical variables. A *non-translational* pragmatics (p. 66) is forthcoming if we assume the underlying semantics to be a non-translational one. (A non-translational one, it will be recalled[12], contains no translations of the object-language expressions and is concerned solely with certain semantical relations between or among them. See VI below.) And a more restricted kind of pragmatics emerges if all semantical notions whatsoever are dropped. In this way we gain a *pragmatics presupposing only a syntax* (p. 68). Such a pragmatics is a good deal more interesting and powerful than one might suspect at first glance.

As goes the pragmatics of acceptance, so goes, *mutatis mutandis*, many another pragmatical relation. Utterance (p. 69) is a such a one. Let

'*X* Utt *a,t*'

express that *X* utters *a* at *t*. This form is essentially Carnap's (5′). Concerning Utt we need a *Restrictive Principle* that only sentences (of some object-languages) are uttered. Also some *Temporal Principles* are needed, that if *X* utters *a* at *t* he or she does not then utter *a* at a proper temporal part of *t*. Also if *X* utters at *t* a sentence containing two or more sentences as segments, then he or she utters the left segment prior to uttering the right one. And in general, the syntactic order of the sentences contained within some compound sentence *a* and the temporal order of utterance are to coincide.

Assertion may be regarded as simultaneous utterance and acceptance. Also relations for believing, knowing, and other epistemic notions may be introduced. And also relations for commanding, exclaiming, and subjunctivizing. The latter, however, were not discussed in *TSP*.

Why, now, we might well ask, are intensions important, more particularly, the purely pragmatical ones? The best answer would seem to be that *parts* of them are of great help in describing logical forms or grammatical deep structures. It has been observed above how all the kinds of pragmatical intensions consist of parts or components, and these are often of more interest than an intension in its totality.

Let us consider an example, say, the sentence

(7) 'Henry admired Kempff's superb playing of the *Waldstein*'.

'Admired' here is to be construed intensionally and 'playing' is to be handled gerundively. The notions discussed in *TSP* do not suffice for a logic of gerundives. But we may forge ahead a little by letting

'X Admire_t e,a'

express that person *X* admires with conscious awareness at time *t* the event (or happening or process or occurrent) *e* as described by the one-place predicate *a* — or, in Frege's phrase, as taken under the *Art des Gegebenseins a*.[13] One rendition of (7) is, then, where 'h' represents Henry, 'k' the great German pianist Wilhelm Kempff, and 'w' the *Waldstein*, as

(7′) ‘(E*e*) (E*a*) (Ee_1) (Ee_2) (h Agent *e* . <Admire> *e* . *e* B now . e_1 Object *e* . e_1 Under$_e$ *a* . *a* = ‘$\hat{e}_2$ (k Agent e_2 . Superb e_2. <Play> e_2 . w Object e_2)’)’.

Here ‘Agent’ represents the quasi-grammatical relation of being an *agent* of, ‘Object’ of being an *object* of, ‘<Admire>’ and ‘<Play>’ are event-descriptive predicates for the gerundives *admiring* and *playing* (in the sense of playing a musical work, as in accord with entry IV. 16 under ‘play’ in the *O.E.D.*), Under is the Frege-like under-relation[14], and ‘*e* B now’ expresses that *e* took place before the now of the present moment.[15] ‘Superb e_2’ will be commented on in a moment.

It might appear that in (7′) we have lost all sight of a purely pragmatical intension. But no, this is not the case. We have here merely a more developed form of one. The very notion ‘Under’, here in the context, may be defined away in terms of our pragmatical stand-by ‘Acpt’. In fact, we may in general let

‘*X* Under$_e$ e_1,*a*’ abbreviate ‘(Per *X* . *X* Agent *e* . PredConOne *a* . (E*b*) (E*t*) (EE_2) (*b* Des e_1 . *e* Throughout$_{\text{Time}}$ *t*. <*X*,Acpt, ($a^\frown b$),*t*> e_2 . *X* ConsAware$_t$$e_2$))’.

Here ‘Per’ represents being a person, ‘PredConOne’ represents being a one-place predicate-constant, ‘*e* Throughout$_{\text{Time}}$ *t*’ expresses that the duration of *e* extends throughout *t*, and ($a^\frown b$) is of course the result of concatenating *a* with some designator *b* of e_1. ‘ConsAware’ is a new predicate needed here to assure that *X* is *consciously aware* of e_2 during *t*. Also

‘<*X*,Acpt,($a^\frown b$),*t*> e_2’ here is merely short for ‘(*X* Agent e_2 . <Acpt> e_2 . ($a^\frown b$) Object e_2 . e_2 Throughout$_{\text{Time}}$ *t*)’.

With these notions in mind, we see that (7′) is equivalent to

‘(E*e*) (E*a*) (Ee_1) (h Agent *e* . <Admire> *e* . *e* B now . e_1 Object *e* . h Under$_e$ e_1,*a* . *a* = ‘$\hat{e}_2$ (k Agent e_2 . Superb e_2 . <Play> e_2 . w Object e_2)’)’.

This is to the effect that Henry is an agent of a past act of admiring some e_1 consciously taken by him under the description of its being a superb playing by Kempff of the *Waldstein*. Or, equivalently, that Henry admired some e_1 such that, where *b* designates e_1, he consciously accepted a sentence to the effect that e_1 is (or was) a superb playing of the *Waldstein* by Kempff.

In this way we see again that Acpt is a most fundamental pragmatical notion, the intensional predicate 'Under' being definable in its terms. It would be premature to say that *all* intensionality required in the study of logical form can be handled in terms of 'Acpt', although it might well turn out that this is the case. The serious study of the logical forms of natural sentences is still in its infancy, most of the forms given to date being mere rough first approximations.

A word now about the adjectival 'superb', which may also be analyzed in terms of 'Acpt'. Let Less-Superb-Than be a comparative relation between entities of a given sort. The entities in the foregoing example are (full) playings of the *Waldstein*. Let

(8) 'e_2 High Less-Superb-Than,W'

express that e_2 is placed high in the series of all playings whatsoever, W, of the *Waldstein* with respect to the Less-Superb-Than relation. The least superb are placed at the bottom of the scale, so to speak, the most superb at the top. We might parse 'Superb e_2' occurring in the modes of description *a* in (7) merely as (8). (If it were not so occurring, some mode of intensionality should be built into (8) itself. Suppose in fact there are no *e*'s of the kind required in (8). In place of 'W' here, we could then require the structural-description of 'W', and then interpret (8) in such a way that the speaker places e_2 high in the Less-Superb-Than scale relative to 'W' even though W itself might contain no superb playings.)

Note that the parsing of (7) as (7′) is such that (7) is "read" so as to require that Henry is self-consciously aware of what he is admiring. This is in accord with the use of 'Under' here. Carnap's 'T' and 'Acpt' are taken so as not necessarily to involve conscious awareness on the part of the agent, however, it will be recalled.

Of course, these are many other "readings" of (7). Suppose Henry listened to a recording but did not know who was playing what, although he did admire what he heard. Or (7) could be taken as expressing that Henry usually admired — or frequently or habitually — admired Kempff's usually, frequently, or habitually superb playing of the *Waldstein*, or that he admired (usually, frequently, and so on) just Kempff's playing of the *Waldstein* that he regards as superb. There are no doubt further readings also, all of which can be handled here on the basis of this kind of material.

The chief value of the pragmatical intensions of *TSP* is thus probably that they lead by easy stages into a theory of logical form or grammatical deep structure. The same cannot be claimed of a theory of purely semantical intensions in which the user of language plays no role.

Let us return now to Carnap's discussion of 1955 in "Meaning and Synonymy in Natural Language," concerning Karl and the predicate 'Hund'. Carnap urged that the extension of 'Hund' for Karl can be determined by a linguist studying Karl's linguistic behavior. In fact, we may accomodate what Carnap says about this in terms of 'Under'. Let 'H' represent the predicate 'Hund' and let 'k' be short for 'Karl'. Then

$$\hat{x}\text{k Under}_t\, x,\text{'H'}$$

is the extension of 'H' for *k* throughout some period *t* of testing, "the class of things to which Karl is willing to apply the predicate;"

$$\hat{x} \sim \text{k Under}_t\, x,\text{'H'}$$

concerns the negation of that extension, where "Karl is not willing to affirm" the predicate;

$$\hat{x}\text{k Under}_t\, x,\text{'-H'},$$

where -H is the Boolean negation of H, is the extension of the negation of the predicate, where Karl "denies" the predicate; and

$$\hat{x}(\sim \text{k Under}_t\, x,\text{'H'} \,.\, \text{k Under}_t\, x, \text{'-H'})$$

is "the intermediate class for which Karl is not willing either to affirm or to deny the predicate." All of these classes are of course relativized to Karl during the time *t*. All of these classes are *subjective* (or pragmatical) extensions of 'H'. And, in similar fashion we can go on to distinguish *intersubjective extensions*, and even *objective* ones. In this way we see that the very notion of an extension, when looked at closely and taking into account the actual users of language and empirical testing procedures, subdivides into many kinds, some of which are pragmatical, not just semantical, in nature.

Carnap's "intensionalist thesis," it will be recalled from *Meaning and Necessity*, pp. 237 ff., is to the effect that the linguist can "assign to one and the same predicate different intensions," there being "more than one and possibly infinitely many properties whose extension within the given region is just the extension determined by the predicate." The difficulty with this, as noted above, is that it utilizes an unanalyzed notion of *property*. But here too, if we bring in pragmatical considerations, we can make headway. Instead of speaking of properties we may bring in varying modes of description or *Arten des Gegebenseins*. Thus, to use Carnap's example, let us consider the intensions of 'horse' ('H') and of 'horse or unicorn' (the Boolean

'(H $\cup$ U)'). Since there are no unicorns, the extensions of these two predicates are the same. Yet a linguist, investigating someone's usage of these phrases, may find that they have very different intensions.

Let us recall Carnap's characterization of being synonymous, that "two expressions are synonymous in the language *L* for [person] *X* at time *t* if they have the same intension in *L* for *X* at *t*." This we may handle here by letting

'a Syn_X b,t' abbreviate '(PredConOne a . PredConOne b . $(x)(X\ \mathrm{Under}_t\ x,a \equiv X\ \mathrm{Under}_t\ x,b)$)'.

Thus two one-place predicate-constants are synonymous for *X* at *t* just where everything *X* takes under the one *X* takes under the other and conversely. Sameness of pragmatical intension seems well captured by this definition. And, then, a subjective pragmatical intension as such, for *X* at *t*, of a predicate *a* is the class of synonymous predicate constants, so that

'int(a,X,t)' is short for '$\hat{b}a\ \mathrm{Syn}_X\ b,t$'.

Clearly we have then that

$$\vdash (a)(b)(a\ \mathrm{Syn}_X\ b,t \equiv \mathrm{int}(a,X,t) = \mathrm{int}(b,X,t)),$$

so that the full content of Carnap's characterization of synonymy is taken care of here by this notion of a pragmatical intension.

Note now also that although the semantical extensions of 'H' for 'horse', and of '(H $\cup$ U)' for 'horse or unicorn', coincide, their pragmatical intensions, for Karl, say, at *t*, need not. As Carnap points out (p. 238), in an empirical test situation the linguist could "simply describe for Karl cases, which he knows to be possible, and leave it open whether there is anything satisfying those descriptions or not. He may, for example, describe a unicorn ... [as] "a thing similar to a horse, but having only one horn in the middle of the forehead". Or he may point to a thing and then describe the intended modification in words, e.g.: "a thing like this one but having one horn in the middle of the forehead". Or, finally, he might just point to a picture representing a unicorn. He then asks Karl whether he is willing to apply the word ... ['horse' or, 'H'] to a thing of this kind." If the answer is affirmative, '(H $\cup$ U)' cannot be taken as a member of int('H',k,*t*), whereas '(H $\cup$ U)' is of course a member of int('(H $\cup$ U)',k,*t*). Thus these two intensions do not have the same membership although H and (H $\cup$ U) have the same semantical extension. Analogous considerations obtain for intersubjective and objective pragmatical intensions, although Carnap does not mention these. It seems very natural, however, in the present context, to do so.

With these various considerations in mind, we see that the material in *TSP* and the somewhat broadened pragmatical notions of extension and intension that grow out of it, carry forward, in a natural way, Carnap's seminal suggestions concerning the foundations of pragmatics, but with added systematicity as regards the exact logical character of the metalanguages needed, and with all manner of additional detail concerning the many kinds of intensions, logical forms, and so on. Pragmatics itself is *semper formanda* with no end in sight.

CHAPTER FOUR

On comparatives: P.T. Geach

"Sic agitur censura et sic exempla parantur:
Quum vindex alios quod monet ipse facit."

In an intriguing recent paper[1], P.T. Geach has pointed out that "the comparative construction is not ... satisfactorily treated by modern linguists or by modern logicians." He calls attention to "intentional comparisons," which "have been considerably neglected by logicians, even in writings on intentional contexts, but are frequent in ordinary language. 'Jones is fatter than he was' might be accomodated to the treatment of 'is fatter than' as an unanalyzed binary relative term by taking it as a comparison between different temporal slices of Jones. But no such treatment will avail for 'Jones looks fatter than he is' or 'Jones is fatter than he looks'."

Geach points out "that logicians have tended to analyze the positive degree in terms of the comparative rather than conversely," largely on the grounds that "the relation answering to a comparative is often much sharper than the concept expressed by the corresponding positive." In this tendency logicians have followed the lead of Carnap, in his seminal discussions of classificatory, comparative, and quantitative concepts.[2] As Carnap points out, "quantitative concepts are no doubt the most effective instruments in the scientific arsenal," but comparative ones have their important role to play also. "Comparative concepts (sometimes called topological or order concepts) serve for the formulation of the result of a comparison in the form of a more-or-less statement without the use of numerical variables."

Let us first reflect upon

(1) 'Jones is fatter than he was',

but without splitting Jones up into temporal slices between expressions for which 'is fatter then' is significant, and without using numerical variables. It is important to note here the two tenses involved. Let us take the 'is' in the present tense — the most natural reading, surely — and parse this sentence in terms of Jones's *states* of being fat. At certain points in all linguistic work state-variables are needed, as well as variables for events, acts, processes, and the like. (The use of such variables with their attendant event-descriptive predicates is the stock-in-trade of process- or event-logic.[3]) Using a more or less familiar notation, let us let

(2) 'Jones High Less-Fat-Than,'HM''

express that the speaker places Jones high in the comparative scale of being less-fat-than as confined to the predicate 'HM' for human males (or for human males more or less like Jones in key respects). Then

'<Jones,High,Less-Fat-Than,'HM'>e'

expresses that e is a state of Jones's being so placed. There is also a relation Higher for being-higher-than as between human states of fatness, so that 'e Higher e', 'HSF'' may express that e is higher-than e' in this sense. Then, During and Before$_{\text{Time}}$ being suitable temporal relations, our sentence (1) may be parsed as

(1′) '(Ee)(Ee')(Ee_1)(Ea)(<Jones,High,Less-Fat-Than,'HM'> e . <he,High,Less-Fat-Than,'HM'> e' . <Assrt> e_1 . e_1 Of$_{\text{Object}}$ a . (1) a . 'he' CrRef$_{\text{a}}$ 'Jones' . e During now . e' Before$_{\text{Time}}$ e . e Higher e','HSF')'.

Here 'now' is of course a deictic expression for the present moment of the speaker, and CrRef is the relation of cross-referentiality. '(1) a' expresses that a is of the form (1). This analysis avoids splitting Jones up into temporal slices; it handles the positive in terms of the comparative — 'Jones is fat' taken timelessly is expressed by (2); it speaks of states of Jones in essentially the same way that we would speak of his actions or occurrences that he undergoes or that happen to him; and so on.

Note that it follows from our sentence (1) or (1′) that Jones is now fat and also that Jones was fat. This is in accord with the "normal" reading of (1). However, (1) could be read in such a way as to allow that Jones is now fat but that he was not then, or even that he is not now fat nor was he then. Suppose he is and always has been thin. Our sentence still makes good sense, but in the logical form the two occurences of 'High' must be replaced by 'Low', where

'Jones Low Less-Fat-Than,'HM''

expresses that Jones is placed low in the Less-Fat-Than scale. (Similarly 'Mid' is used for the form to express that he is placed somewhere in the middle, and the forms containing 'High', 'Mid', and 'Low' are then understood to be mutually exclusive and jointly exhaustive.) And, similarly, where Jones is now fat but was not then, only the second occurrence of 'High' is replaced by 'Low'. Another reading is where we do not wish it to follow that Jones either is or ever was either fat or thin. In this case we can use 'Mid' in the logical form. If (1) is read so as not to have any consequences about his being fat or thin or either, a kind of disjunction can be used. For this, let

> '<Jones,(High ∪ Mid ∪ Low),Less-Fat-Than,'HM'> *e*' abbreviate '(<Jones,High,Less-Fat-Than> *e* v <Jones,Mid,Less-Fat-Than,'HM'> *e* v <Jones,Low,Less-Fat-Than,'HM'> *e*)'.

Our form would then be (1′) with '(High ∪ Mid ∪ Low)' replacing 'High' both times.

Can we make the foregoing sort of analysis work also for the kinds of intentional sentences to which Geach calls attention? Let us try. Consider first

(3) 'Jones looks fatter than he is'.

This may be given the form

(3′) '(E*e*)(E*e*′)(E*a*)(E*e*′′)(Ee_1)(E*b*)(Look *e* . *e* During now . *e* $\text{Of}_{\text{Object}}$ *e*′ . *e*′ St Jones . *e*′ During *e* . *e*′ Under_e *a* . '{e_1 ∋ <Jones,(High ∪ Mid ∪ Low),Less-Fat-Than,'HM'> e_1}' *a* . <he,(High ∪ Mid ∪ Low),Less-Fat-Than,'HM'> *e*′′ . <Assrt> e_1 . e_1 $\text{Of}_{\text{Object}}$ *b* . (3) *b* . 'he' CrRef_b 'Jones' . *e*′′ During now . (e_2)((e_2 During now . e_2 Under_e *a*) ⊃ e_2 Higher *e*′′,'HSF'))'.

Here *e* is a present look of a state *e*′ of Jones under the description of *e*′'s being a state of Jones's being fat or not, *e*′′ is a present state of Jones's being far or not, and every e_2 taken in *e* under that description is higher-for-human-fatness-states than *e*′. The Under-relation here is that used in I and II above. 'Look *e*' expresses that *e* is a look (or result of an act of looking at), $\text{Of}_{\text{Object}}$ is the prespositional relation of being the object-of, St is the relation of being a state of, and "{*e* ∋ ---*e*- --}' *a*' expresses that *a* is an inscription of the given virtual-class form. From (3) it does not follow that Jones is fat. Jones might be thin but still look fatter than he is.

Geach is surely correct in objecting to the transformational grammarian approach to

(4) 'Tom is taller than Bill is'

in terms of transformations of some sort from 'Tom is tall' and 'Bill is tall'. He says that he does "not see how anything like a function from senses to senses can get us from the senses of these two sentences to the sense of the comparison, *and* will give the right result also in other cases." Quite. (4) is a relational sentence, for the handling of which a relational term is needed *ab initio*.

Geach comments that "an even more objectionable bit of theorizing that I have come across in some linguists is explanation of comparatives by quantifying over degrees. For how are we to analyze (say) 'A is brighter than B' in terms of degrees of brightness? If the possession of a certain degree of brightness by A or by B excludes the possession by the something of any other degree of brightness, then presumably the analysis will be 'The degree of brightness of A is higher than the degree of brightness B has'; but 'higher than' at least looks like one more example of the comparative degree we are trying to explain; this is not encouraging." No, it is not. Still, 'higher than' could be construed as the '>' of real-number theory, say, so that we could presumably have that

$$\text{A is brighter than B} \equiv (\text{E}\alpha)(\text{E}\beta)(\text{NR}\ \alpha\ .\ \text{NR}\ \beta\ .\ \alpha = \text{the degree of brightness of A}\ .\ \beta = \text{the degree of brightness of B}\ .\ \alpha > \beta).$$

This kind of rendition might be suitable for some types of "things" A and B. Still, Geach is surely correct in not wishing to force this kind of analysis on all comparative terms.

Geach comments that "it would be agreeable if comparisons could be all analyzed in terms of the relation *greater than* between real numbers, ...; but I fear this is impossible, in particular for intentional comparisons. A comparative adjective often gives expression to a clear method of rank ordering when we have no means of measurement with real numbers; and even when we have, A may believe that B is more so-and-so than C although A is thoroughly innumerate and does not believe anything to do with real numbers." This last contention seems to rest on a confusion. *We* can parse a sentence of the form 'A believes B is more so-and-so than C' without ascribing any knowledge on A's part with the way in which we parse 'B is more so-and-so than C'. Consider

‘Peter believes Jones is fatter than Smith’,

or, better,

(5) ‘Peter believes Jones is fatter than he (actually) was’,

the ‘(actually)’ here serving only to assure the cross-referentiality of ‘he’ to ‘Jones’. We — we logico-linguists — can give this the form

(5′) ‘(Ee)(Ea)(<Blv> e . Peter Agent e . e Of$_{\text{Object}}$ a . a That$_{\text{Content}}$ (1′))’,

where That$_{\text{Content}}$ is the that-relation of content, ‘<Blv>’ the gerundive for believing, and Agent the relation of being the agent of. In place of ‘(1′)’ here, of course, the formula above thus labelled is to be inserted. In citing a form — or in “deriving” it by transformation rules from the parent English sentence[4] — we do not ascribe any knowledge to Peter concerning the logical form (1′) nor indeed concerning (5). Peter may be thoroughly ignorant of all the logical and metalogical theory involved.

We must be careful not to confuse an “explanation” of “belief by innumerate A that B is fatter than B really is” with citing a logical form for a statement of such belief. The two are very different enterprises. To give a logical form is to give a meaning to the sentence. To explain a sentence, for Geach, is apparently to question A and his friends and acquaintances in such a way as to gain suitable grounds or evidence for the given sentence.

Geach gives an analysis of

(6) ‘Jones has become stupider than Jones used to be’.

“in jargon that is (near enough) Fregean.” Let us avoid all such jargon here in favor of the clear-cut notions of event-logic. In order to “scratch only where it itches” and to keep out irrelevancies, let us construe ‘used to be’ just as ‘was’. Recalling (1′) above, we see that a form for (6) can be something like

(6′) ‘(Ee_1)(Ee_2)(Ee)(<Jones,Become,e_2> e_1 . e_1 Before$_{\text{Time}}$ now . e_1 Result e_2 . e_2 During now . <Jones,(High ∪ Mid ∪ Low),Less-Stupid-Than,‘HM’> e_2 . <Jones, (High ∪ Mid ∪ Low),Less-Stupid-Than,‘HM’> e . e Before$_{\text{Time}}$ now . e_2 Higher e,‘HS’))’.

We have had to scratch here to provide for the present perfect aspect of ‘become’ by bringing in the *result* e_2 of the becoming, and stipulating that it is to be a state now of Jones’ being stupid or not higher than his past state of being stupid or not. Here too there are other possible readings.

Geach invites us to reflect upon the validity of a "comparative categorial syllogism" such as

> 'Smith is stupider than Jones has become,
> Jones has become stupider than Jones used to be,
> *Ergo*: Smith is stupider than Jones used to be.'

It is clear from the above how these sentences may be formulated, the validity of this inference then being seen to be merely a special case of validity in first-order quantification-theory.

After his first American tour, the great pianist Artur Schnabel commented:

(7) 'The greater the music, the lower the fee.'

Here there are two deleted or zeroed 'is's, one at the end of each clause, and (7) is no doubt intended as

(7′) 'The greater the performance of great music is, the lower the fee for it is.'

How now may this kind of comparative sentence be handled? Well, we may fix the non-logical vocabulary as follows. Let '*e* Prfm *a*' express that *e* is a mereological sum of performances of a sum of scores of musical compositions, 'SGM *a*', that *a* is a sum of scores of great musical compositions, and '*x* Fee *e*', that *x* is the fee paid for such an *e*. Also 'e_1 Gr e_2,MP' is to express that e_1 is greater-than e_2 where e_1 and e_2 are both musical performances or sums of such, and 'x_1 Gr x_2,F' expresses that x_1 is greater than x_2 is where x_1 and x_2 are fees. A first approximation to (7) in the "reading" (7′) is then

(7″) '$(e_1)(e_2)(a)(x_1)(x_2)((e_1$ Prfm a . e_2 Prfm a . $SGM\ a$. x_1Fee e_1 . x_2 Fee $e_2) \supset (e_1$ Gr e_2,MP $\supset x_2$ Gr x_1,F))'.

Given any two performances of the same sum of scores of great music, if one is greater than the other in the appropriate sense, then the fee paid for it is less than the fee paid for the other.

The difficulty with (7″), however, is that it does not capture the *proportionality* that is no doubt intended. The difference in greatness in the performances is presumably intended to be inversely proportional to the difference in the respective fees. Another reading is needed to capture this. Let 'e_1 VMGr e_2,MP' express that e_1 is very much greater than e_2 as MP's, 'e_1 VVMGr e_2,MP', that e_1 is very very much greater than e_2 is, and so on, as far as one wishes to distinguish.[5] And similarly for 'x_1 VMGr x_2,F', and so on, for the fees. The desired form then becomes

(7‴) '$(e_1)(e_2)(a)(x_1)(x_2)((e_1$Prfm a . e_2 Prfm a . SGM a . x_1 Fee e_1 . x_2 Fee $e_2) \supset ((e_1$ Gr e_2,MP $\supset x_2$ Gr x_1,F). (e_1 VMGr e_2,MP $\supset x_2$ VMGr x_1,F) . ...))'.

Any further readings needed may no doubt be captured by suitable adjustments in the details.

Along with comparative categorical syllogisms, Geach calls attention to comparative hypothetical syllogisms, such as

'The merrier our party becomes, the noisier our songs will be;
The noisier our songs are, the madder the neighbors will get;
Ergo: the merrier our party becomes, the madder the neighbors will get.'

Geach comments that his "hunch" here is that the validity of "hypothetical syllogisms in the propositional calculus" is "involved" in the validity of syllogisms of this kind. Geach leaves the problem of justifiying such syllogisms "as an exercise to the reader," obviously without realizing how complicated these sentences are when looked at from the point of view of deep structure *from close to*. Oversimplifying somewhat, we have quantificational structures of the forms

$$\ulcorner(e_1)(e_2)(e_3)(e_4)((\boldsymbol{A}_1 . \boldsymbol{A}_2) \supset (\boldsymbol{B}_1 \supset \boldsymbol{B}_2))\urcorner,$$
$$\ulcorner(e_3)(e_4)(e_5)(e_6)((\boldsymbol{A}_2 . \boldsymbol{A}_3) \supset (\boldsymbol{B}_2 \supset \boldsymbol{B}_3))\urcorner,$$

and

$$\ulcorner(e_1)(e_2)(e_5)(e_6)((\boldsymbol{A}_1 . \boldsymbol{A}_3) \supset (\boldsymbol{B}_1 \supset \boldsymbol{B}_3))\urcorner$$

for the two premisses and conclusion, respectively, with suitable restrictions on how the variables can occur in $\boldsymbol{A}_1$, $\boldsymbol{A}_2$, etc. The inference, to be valid, requires a tacit existence assumption,

$$\ulcorner(\mathrm{E}e_3)(\mathrm{E}e_4)(\boldsymbol{A}_2 . \boldsymbol{B}_2)\urcorner,$$

roughly, that at least one song at our party is noisier than at least one other. Geach's "hunch" here is thus in a way correct but too simple to cover the full quantificational character of the inference involved.

Geach cites another syllogism he apparently takes to be of the same form, namely,

'The uglier a woman is, the more cosmetics she uses;

> The more cosmetics a woman uses, the less ugly she looks;
> *Ergo*: the uglier a woman is, the less ugly she looks.'

He comments that this syllogism is invalid, but that "it would be following a false clue to look for the cause of invalidity in the intentionality of 'the less ugly she looks'; for the syllogism remains invalid if we replace this with 'the fewer people are repelled by her' both times." But 'repelled by' is surely to be handled intentionally, as well no doubt as 'ugly' is. Appeal to the intentional character of these notions would seem to be essential here.

At the end of his paper, Geach considers "the logically disconcerting effect of having two comparatives or two superlatives together." He worries, for example, about the comparison involved in the statement of a "policy designed for the greatest happiness of the greatest number, i.e. intended to make *more* people *happier* than any other policy. Are we sure we understand the logical import of this double comparison? I [Geach] am inclined to say that there is nothing to understand. 'The greatest happiness of the greatest number' is surely an immensely effective bit of rhetoric, like Churchill's 'Never was so much owed to so many by so few,' and if we take it just as that it is idle to seek to milk out a precise sense." Well, even a rhetorical sentence has its logical form, namely, the logical form of an appropriate, non-rhetorical, paraphrastic sentence. Of course 'the greatest happiness of the greatest number' is meaningless in isolation, but can be given meaning within the context of a theory, or within some suitable sublanguage of a natural language. One way would be as follows.

Let 'policy A' express that A is a policy, i.e., a "text", and let '*b*' be a variable over sentences, texts, and the like, just as '*a*' has been above. Let 'SG *p*' express that person *p* is a member of the social group under discussion, '*p* Able_t *e*,'MHH'' to express that person *p* is able or capable at time *t* of being in a state *e* as taken under the description of its being a state of maximal human happiness, '*e* Accord A' that *e* is a state in accord with *A*, '*p* St_t *e*' that person *p* is in the state *e* at time *t*, and '*t* Approp *p*' that *t* is a time appropriate to or during the life-span of *p*. Happiness here is not taken as something external to the person, but as something he or she must in part work for and may or may not be able of achieving maximally. And some policies might be such as to exclude this maximal state for many. Consider now the form

> '$((p)(e)(t)((\text{SG}\ p\ .\ p\ \text{Able}_t\ e,\text{'MHH'}\ .\ t\ \text{Approp}\ p) \supset (e\ \text{Accord A}\ .\ p\ \text{St}_t\ e))\ .\ \sim (\text{E}b)\ (\text{Policy}\ b\ .\ \sim \text{A} = b\ .\ (p)(e)(t)((\text{SG}\ p\ .\ p\ \text{Able}_t\ e,\text{'MHH'}\ .\ t\ \text{Approp}\ p) \supset (e\ \text{Accord}\ b\ .\ e\ \text{St}_t\ p))))$',

expressing that every person is always in the state of maximal happiness of which he or she is capable at the time, this state being in accord with the policy A, but that this holds of no other policy. 'The greatest number' here causes no problem, designating the number of members of the social group SG, and is paraphrased away. And, of course, the hedonics of the theory is such that 'MHH' for the states of maximal human happiness is a meaningful notion. Rhetorical devices, along with extended usage, metaphors[6], fictional locutions[7], and so on, have a logic all their own, which we must work hard to get at.

How fares now

(8) 'Never was so much owed to so many by so few'?

Here we must take seriously 'much', 'many', and 'few' with all the slack and "fuzziness" thereunto appertaining. Also the owing here is to be handled by a term taken collectively rather than distributively. To handle terms taken collectively[8] it is convenient, again, to bring in mereology or the calculus of individuals.

Let $(p_1 \cup p_2 \cup \ldots \cup p_n)$ be the mereological sum of persons $p_1, \ldots, p_n$. Any such sum is a *fusion of persons*. We may let, in general,

'FuPer p' abbreviate '((Eq)(Per q . q P p) . (x)(($\sim x$ = N . x P p) ⊃ (Eq)(Ey)(Per q . q P p . y P q . y P x)))'

expressing that p is a fusion of persons in the sense that p has at least one person as a part and is such that every non-null part of p has a part that is a part of some person that is a part of p. And similarly 'FuItComp x' may be defined to express that x is a fusion of the relevant *items of compensation*, i.e., of whatever is owed to the many by the few. Let 'p Many Per' express that p is a fusion of many persons, 'q Few Per' that q is a fusion of a few persons, and 'x Much ItComp' that x is a fusion of many items of compensation (i.e., x is much compensation). Also we need comparatives 'More' and 'Fewer' such that 'p More q' expresses that the fusion of person p has more persons as parts than q has, 'p Fewer q', fewer, and 'y More x', that the FuItComp contains more items of compensation than the FuItComp x does. Actually here we may let

'p More q' abbreviate '(FuPer p . FuPer q . q P p . $\sim p$ P q)',

and similarly for 'y More x'. Also we may perhaps define

'p Few Per' as '(Eq)(Er)(q Many Per . FuPer p . $\sim p$ Many Per . (FuPer r v r = N) . (s)(Per s ⊃ s P ($q \cup p \cup r$)))', so that a few added to a many, added perhaps to some other FuPer, is the totality. Owing is to be handled tetradically so that 'p Owe_t x,q' expresses that p owes x to q at time t, where p and q are either separate Per's or mereological sums of such and x is a separate ItComp or a sum of such. With all this, a form for (8) may be given as

(8′) '~ (Et)(Ep)(Eq)(Ex)(FuPer p . FuPer q . FuItComp x . p Many Per . q Few Per . x Much ItComp . t $\mathrm{Before}_{\mathrm{Time}}$ now . p Owe_t x,q . ~ (Ep')(Eq')(Ex')(p' More p . q' Fewer q . x' More x . p' Owe_t q',x))'.

All of these forms are put forward tentatively and as first approximations subject to further refinement and improvement. The serious study of logical form, and of the logical semiotics needed for it, is still in its infancy, and much more thorough work than has yet been undertaken lies ahead. In the foregoing, the non-logical vocabulary used has been *ad hoc* for each sentence. Eventually, of course, we should hope to have a full primitive vocabulary, together with a non-circular lexicon, that would enable us to handle the full complexity of all sentences of the language. Recent advances in lexicography are such as to suggest that this hope is not a mere dream, but a goal to which we are coming closer day by day.

A word now about St. Anselm's famous proof for the existence of God in the *Proslogion*. Geach gives no logical forms in terms of which the argument — whatever it is, and this is by no means easy to say — can be discussed in any logical detail. He thinks that the real "heart of the matter ... lies in the problem of intentional comparisons." Well, in order to make sense of the proof, we must surely distinguish between being an entity "greater than can be conceived" and just being an entity greater than all others. The gist of the argument is to show how to get from a suitable conception of an "unsurpassable" being — to use Hartshorne's word for it — to the existence of such a being. A logical analysis of such a proof has been given previously.[9] The basic non-logical vocabulary needed is an intentional predicate 'Cncv' for *conceiving*, a predicate 'Able' for human *ability* (to conceive), and 'Gr' for a very general, Anselmic relation of being *greater than*. We can then frame premisses in accord with those of Anselm, reflect upon their soundness, and proceed to the conclusion. Nothing here depends upon any problems of intentional comparison, although of course the intentional predicates 'Able' and 'Cncv' and the comparative one 'Gr' play a fundamental role. Here, again, Geach's view is too simple.

To get at an adequate logic of comparatives, we see that nothing can take the place of unearthing the deep logical structures of the sentences as used in a given context. For this, it is essential to press for the full non-logical vocabulary needed and for the quantificational structure, or its surrogate in some sort of relational algebra, combinatory or otherwise, as, for example, in XIII below. Surface grammar is of little help here, except as a heuristic leading to problems at a deeper level. We are all much indebted to Geach for his often brilliant insights and for his depth of understanding of some aspects of the great tradition of the history of logic, and for the breadth and scope of his general philosophic outlook. We can only express regret that he has not seen fit to pay greater heed to the inner technicalia of modern logical semiotics, which would provide him tools more helpful for his work than those currently available in England.

CHAPTER FIVE

On homogeneous logic, a quasi-Leśniewskian ontology, and type-theory

"On revient toujours à ses premiers amours."

The very idea of a homogeneous logic is an attractive one. Everything whatsoever is construed as an entity of one and the same general logical kind or type.[1] Such differences among entities as are to be distinguished are then differences to be accomodated *within* the theory. We live in one unified universe, after all, and everything whatsoever is contained therein. Whatever there is is thus like whatever there is, in just this sense; the entities of the homogeneous theory are just the entities that there are. Such entities we usually call 'individuals' but no philosophic profundity is to be attached to the term. It is merely a convenient rubric for the most fundamental entities admitted, all other entities then being constructs in terms of them. The individuals are the most basic, all other kinds of entities being constructible in terms of them.

Let us reflect upon the extended or ordinal mereology OM, of the previous paper, by way of logical background. The net effect of the discussion here will be a reinterpretation of that system along somewhat more traditional lines. Let us recall the two fundamental primitives, 'P' for a dyadic relation of *part to whole* and 'Ord' for a triadic relation of *ordination*, so that 'x Ord y,z' expresses that x is the *ordinal individual* consisting of y and z in this order. In addition, an operation of *unit summation* is admitted so that '(x1--x--)' is regarded as a term for the sum of unit individuals such that --x--, where '--x--' is some sentential form of the system perhaps containing 'x' as a free variable. What now, we should ask, is a unit individual? One construal is that of EM; another construal is now to be given.

Let us refer hereafter to the system here as HL.

In OM parts of individuals were admitted, including the null individual N. Here in HL an individual will have no parts — in the sense of 'P' — other than itself and the null individual. Recall the definition of

D1. 'N' as short for '$(x1 \sim x \text{ P } x)$'.

Then we may also let

D2. 'Unit x' be short for '$(\sim x \text{ P N} \,.\, (y)((\sim y \text{ P N} \,.\, y \text{ P } x) \supset x \text{ P } y))$',

so that an entity x is a *unit* just where it is not a part of N but is a part of all non-null parts of it. This definition allows for both cardinal and ordinal units, as in EM. Here in HL, however, these latter are to be handled a little differently.

Let now

D3. 'OrdEnt x' abbreviate '$(\text{E}y)(\text{E}z)x \text{ Ord } y,z$',

so that an *ordinal entity* x is any result of ordinating any entities y and z. A *cardinal* entity is then one of all of whose unit parts are non-ordinal entities, so that

D4. 'CardEnt x' may abbreviate '$x \text{ P } (y1 \sim \text{OrdEnt } y)$'.

The notion of a *cardinal individual* can now be construed, within the theory, as any unit, cardinal entity. Thus

D5. 'CardInd x' now abbreviates '$(\text{Unit } x \,.\, \text{CardEnt } x)$'.

And similarly

D7. '$\text{OrdEnt(Card)}_n\ x$' may abbreviate '$(\text{E}y_1)\ldots(\text{E}y_n)(x \text{ Ord } y_1,\ldots,y_n \,.\, \text{CardInd } y_1 \,.\, \ldots \,.\, \text{CardInd } y_n)$',

where

D6a. '$x \text{ Ord } y,z,w$' abbreviates '$(\text{E}u)(u \text{ Ord } z,w \,.\, x \text{ Ord } y,u)$',

and so on, as in OM. An *ordinal entity of degree n whose ordinands are cardinal individuals* is thus now any result of ordinating any n cardinal individuals.

The unit cardinal entities, or cardinal individuals, are to be understood more or less as the "individuals" or "things" or "objects" of everyday life, just as the *Urelemente* of a set-theory may be. Ordinal entities, on the other hand, are something quite new and need not be regarded as "individuals" in any ordinary sense; hence we do not refer to them here as individuals.

By a *collective* let us mean now any sum $(x1\text{--}x\text{--})$ of unit entities x such that --*x*--. Such a sum may be of either cardinal individuals, of ordinal entities, of further kinds of entities to be distinguished, or of any of these kinds of entities summed or ordinated in any way. For the moment, let us reflect upon just collectives of cardinal individuals, i.e., upon sums designated by expressions of the form '(*x*1(CardEnt *x* . --*x*--))'. This notion may be defined by letting

D8. 'Coll(Card) *x*' abbreviate '(*y*)((Unit *y* . *y* P *x*) ⊃ CardInd *y*)'.

This definition is such as to allow for both null and universal cardinal collectives. Of course, one might wish to exclude a null collective on the Leśniewskian ground that there can be no collective if there is nothing to be collected. Here, note, there is no null cardinal individual, a null cardinal collective being admitted in its stead. Leśniewski apparently has it just the other way around — a null individual but no null collective — as we shall see in a moment.

Relations are also to be handled here as collectives, *relational collectives*. And of course there are many different kinds, depending upon the kinds of relata. First let us consider must relational collectives whose relata are cardinal individuals. For this we may let

D9. 'RelColl(Card)_n *x*' abbreviate '(*y*)((Unit *y* . *y* P *x*) ⊃ OrdEnt(Card)_n *y*)'.

An entity *x* is thus a relational collective of degree *n* with only cardinal entities as ordinands just where every unit part of *x* is an ordinal entity of degree *n* whose ordinands are cardinals. Relational collectives (among cardinal entities) are thus built up here as sums of ordinal entities just as cardinal collectives are built up as sums of cardinal individuals.

By a sum, of course, throughout the discussion, is meant merely a Boolean sum. The null individual has already been introduced as the Boolean sum of all unit entities *x* such that ~ *x* P *x*. The other Boolean notions are immediately forthcoming in full generality as follows.

D10a. '(*x* ∪ *y*)' as '(*z*1(*z* P *x* v *z* P *y*))',
D10b. '(*x* ∩ *y*)' as '(*z*1(*z* P *x* . *z* P *y*))',
D10c. '-*x*' as '(*y*1 ~ *y* P *x*)',

and

D10d. 'U' as '-N'.

These definitions introduce the Boolean sum of x and y, the Boolean product of x and y, the negative of x or anti-x, and the universal or cosmic element.

As basic principles of the system, the following are needed, essentially as in OM. Clearly P is totally reflexive and transitive so that

Pr1. $\vdash (x) x \mathrm{~P}~ x$,
Pr2. $\vdash (x)(y)(z)((x \mathrm{~P}~ y \,.\, y \mathrm{~P}~ z) \supset x \mathrm{~P}~ z)$.

Also we have *Principles of Unit Summation* to the effect that

Pr3. $\vdash (y)((x1\text{--}x\text{--}) \mathrm{~P}~ y \equiv (x)((\mathrm{Unit}~x \,.\, \text{--}x\text{--}) \supset x \mathrm{~P}~ y))$,
Pr4. $\vdash (y)(y \mathrm{~P}~ (x1\text{--}x\text{--}) \equiv (x)((\mathrm{Unit}~x \,.\, x \mathrm{~P}~ y) \supset \text{--}x\text{--}))$,

where '--*x*--' is any sentential function perhaps containing '*x*' as a free variable.

Concerning Ord we have the following basic principles.

Pr5a. $\vdash (x)(y)(z)(w)((x \mathrm{~Ord}~ y,z \,.\, w \mathrm{~Ord}~ y,z) \supset x \mathrm{~P}~ w)$,
Pr5b. $\vdash (x)(y)(z)(w)((x \mathrm{~Ord}~ y,z \,.\, x \mathrm{~Ord}~ w,z) \supset y \mathrm{~P}~ w)$,
Pr5c. $\vdash (x)(y)(z)(w)((x \mathrm{~Ord}~ y,z \,.\, x \mathrm{~Ord}~ y,w) \supset z \mathrm{~P}~ w)$.

These principles stipulate in effect that Ord is a one-many-many, a many-one-many, and a many-many-one relation in the sense appropriate for triadic relations.

Identity may be regarded as merely mutual part-to-whole, so that

D11. '$x \mathrm{~Id}~ y$' abbreviates '$(x \mathrm{~P}~ y \,.\, y \mathrm{~P}~ x)$'.

We write 'Id' here rather than the more usual '=', in order to introduce this latter later (*DL5*) in its strict Leśniewskian sense.

With 'Id' available, we note that *Pr5a-5c* also now hold with 'Id' replacing 'P' in the consequents. Also we have additional principles of identity as follows.

Pr6. $\vdash (x)(y)(x \mathrm{~Id}~ y \supset (z)(u)((x \mathrm{~Ord}~ z,u \equiv y \mathrm{~Ord}~ z,u) \,.\, (z \mathrm{~Ord}~ x,u \equiv z \mathrm{~Ord}~ y,u) \,.\, (z \mathrm{~Ord}~ u,x \equiv z \mathrm{~Ord}~ u,y)))$,
Pr7. $\vdash (x)(y)(z)(w)(u)(v)((x \mathrm{~Ord}~ y,z \,.\, w \mathrm{~Ord}~ u,v) \supset (x \mathrm{~Id}~ w \equiv ((y \mathrm{~Id}~ u \,.\, z \mathrm{~Id}~ v) \vee (\mathrm{E}t)(y \mathrm{~Ord}~ u,t \,.\, v \mathrm{~Ord}~ t,z) \vee (\mathrm{E}t)(z \mathrm{~Ord}~ t,v \,.\, u \mathrm{~Ord}~ y,t))))$.

The following special principles concerning Ord and OrdEnt are also needed.

Pr8. $\vdash (x)(y)(z)(x \mathrm{~Ord}~ y,z \supset (\sim x \mathrm{~P}~ y \,.\, \sim x \mathrm{~P}~ z))$,
Pr9. $\vdash (x)(y)((\mathrm{OrdEnt}~x \,.\, \sim \mathrm{P~N} \,.\, y \mathrm{~P}~ x) \supset x \mathrm{~P}~ y)$,

Pr10. $\vdash (x)(y)(z)(x \text{ Ord } y,z \supset ((y \text{ P } x \supset y \text{ P N}) \,.\, (z \text{ P } x \supset z \text{ P N})))$.

These principles are to the effect that every ordinal entity is a part of all of its non-null parts, and that if x is an ordinal entity with y and z as its ordinands, x is not a part of either y or z, and y and z are not parts of x unless they are null.

Thus far nothing has been said concerning *Principles of Existence*. We assume first an *Axiom of Infinity for Ordinal Entities*, to the effect that, given any entities x and y that lack a non-null part in common, there always exists an ordinal entity with them as ordinands. (This assumption is very strong, perhaps too strong in some respects. Cf. *Pr8* and *Pr9 inter alia* in the Appendix below.)

Pr11. $\vdash (x)(y)((\text{E}z)(\sim z \text{ P N} \,.\sim z \text{ P } x \,.\sim z \text{ P } y) \supset (\text{E}z)z \text{ Ord } x,y)$.

Clearly there could not be an ordinal entity y with U as one of its ordinands, for then y itself would have to be a part of one of its ordinands — which would violate *Pr9*, requiring (together with *Pr8*) that every ordinal entity be a unit. And similarly if the sum of x and y were universal. In a moment another principle of existence will be given, an *Axiom of Infinity for Cardinal Individuals*.

Let us compare now — albeit very roughly — the foregoing framework with Leśniewski's so-called Ontology, the central part no doubt of his trivium of Protothetic, Ontology, and Mereology. A full understanding of Leśniewski's work is not easy to come by, and most of us must rely upon secondary sources for such meager understanding as we have. For the present purposes, we shall rely on the valuable exposition of Czeslaw Lejewski, which is generally regarded as a highly trustworthy and reliable one.[2] In Ontology the concern is with individuals and collectives, and relations are somehow to be provided for, it would seem, in terms of these.

Lejewski, following Woodger,[3] speaks of "shared", "unshared," and "fictitious" names. Shared names are taken to name more than one object; unshared names name one and only one object; and fictitious names are "expressions which as regards their syntax behave like shared or unshared names, but which do not name anything, *e.g.* 'Pegasus', 'centaur', 'object which does not exist'." Lejewski comments (p. 153) that Leśniewski's "Ontology has been described as the most comprehensive logic of Names because its most characteristic expressions belong to the semantical category of names, ... If, however, we take into account the contents of Ontology then it would be more appropriate to describe it as a theory of what there is. Just as astronomy tells us something about heavenly bodies, the theses [or principles] of Ontology tell us something about things, or objects if one prefers, or individuals."

Very well, let us take Ontology as a theory of individuals — and of how they are taken together to form collectives — and take its variables as ranging over such. Lejewski, however, writes in some passages as though the variables are to be taken as "standing for names." Thus, he explains that Leśniewski's 'ε' or "functor of singular inclusion" is allowed to occur "in expressions of the form '*a* ε *b*' (to be read: *a* is *b*)." A sentence of this form, we are told, is true just where "the semantical status for which the '*a*' and the '*b*' stand" is such that '*a*' is an unshared name, '*b*' is a shared or unshared name, and any individual designated or denoted by '*a*' "is" or "is an" — in the sense of Leśniewski's 'ε' — individual denoted by '*b*'; otherwise false. This sentence is a paraphrase of Lejewski's and seeks to capture its intent in the exact terms of the semantics of designation and denotation. Lejewski's explanation apparently fails to distinguish the use and mention of '*a*' and '*b*'. On the one hand, these expressions "stand for" names, and, on the other, '*a* ε *b*' is read: *a* is *b* — as though '*a*' and '*b*' stood for objects, not names.

For present purposes, let us think of a shared name as designating a non-null, non-unit cardinal collective (in the sense of *D8* above), an unshared name as an individual constant designating an individual uniquely, and a fictitious name as designating either the null individual N or a null collective in the sense of *D8*. It was commented above that collectives in the sense of *D8* are allowed to be null, and also they are allowed to contain a single individual. A Leśniewskian collective, however, cannot be null and cannot contain a single individual. To provide for this notion, we merely let

D12. 'LesColl(Card) *x*' abbreviate '(Coll(Card) *x* . ~ *x* P N . ~ CardInd *x*)'.

Also for fictitious entities, we may let

D13. 'Fict *x*' abbreviate '*x* P N'.

The question now arises as to the extent to which the present framework contains the full Leśniewskian theory as sketched by Lejewski. To determine this let us refect upon Leśniewski's various "functors" of *inclusion, exclusion, identity*, and *existence* and see how they may be provided for here. We will then be in a position to see how they may be interrelated and the various Leśniewskian Axioms of Ontology accomodated.

The following definitions, if not actually capturing the *fine pointe* of Leśniewski's intent, would seem to come very close.

DL1. 'x ε y' for '((CardInd x . CardInd y . x Id y) v (CardInd x . LesColl(Card) y . x P y))',

DL2. '$x \sqsubset y$' for 'x ε y v (LesColl(Card) x . LesColl(Card) y . (x Id y v x P y)))',

DL3. '$x \subset y$' for '($x \sqsubset y$ v (Fict x . CardInd y . ~ x Id y) v (Fict x . LesColl(Card) y . ~ x Id y) v (Fict x . Fict y . ~ x Id y))',

DL4. '$x \triangle y$' for '($x \sqsubset y$ v (LesColl(Card) x . CardInd y . y P x) v (LesColl(Card) x . LesColl(Card) y . y P x) v (LesColl(Card) x . LesColl(Card) y . (Ez)(LesColl(Card) z . z P x . z P y)))'.

The four relational expressions or "functors" introduced in these definienda are for *singular inclusion, strong inclusion, weak inclusion,* and *partial inclusion*, respectively. The "functors of exclusion" are the dyadic-relational negatives of these functors.

The three "functors of identity" are as follows.

DL5. '$x = y$' for '(CardInd x . CardInd y . x Id y)',

DL6. '$x \square y$' for '($x = y$ v (LesColl(Card) x . LesColl(Card) y . x Id y))',

DL7. 'x 0 y' for '($x \square y$ v (Fict x . Fict y . x Id y))'.

These definienda introduce respectively the functors of *singular* identity, *strong* identity, and *weak* identity. The "functors of existence" are introduced contextually as follows.

DL8. 'Ex x' for '(CardInd x v LesColl(Card) x)',

DL9. 'Sol x' for '(CardInd x v Fict x)',

DL10. 'Ob x' for 'CardInd x'.

The definienda are respectively, to the effect that x *exists* or that there is *at least one* x, that there exists *at most one* x, and that there exists *exactly one* x.

Lejewski lists several "theses" concerning these various notions, which can be shown to hold on the basis of the principles of HL given above. He also lists several different single Axioms of Ontology, or axiom sets, that have been shown to suffice for Leśniewski's theory. Presumably all of these can be shown to hold here within HL also. It thus seems that Leśniewski's Ontology, as developed by Lejewski, can be mirrored here. The resources of HL, however, as is already evident, differ radically from those of Leśniewski in the handling of relations. Relations are not regarded here as constituting some new "semantical category," but are built up from ordinal entities in essentially the same way that cardinal collectives are built up from cardinal individuals. Let us reflect now upon how this may be done in such a way as to capture the full mathematical power of the simplified theory of types. First, however, a word about the arithmetic of finite cardinals.

Let us let

D13. 'x Suc y' abbreviate '(Coll(Card) x . Coll(Card) y . (Ez)(CardInd z . $\sim z$ P y . $x = (y \cup z)$)',

expressing that the cardinal collective x is a *successor* of the cardinal collective y in the sense of being a sum of y with a cardinal individual not a part of y. We postulate now that every cardinal collective has a (cardinal-collective) successor. This is in effect an *Axiom of Infinity for Cardinal Individuals*.

Pr.12 $\vdash$(x)(Coll(Card) $x \supset$ (Ey)(y) Suc x).

It then immediately follows that

$\vdash$(Ex)CardInd x.

How now can we handle collectives of collectives of cardinal individuals? We cannot merely form sums of them for the summands would then become "lost" in the sum, so to speak. In a sum of cardinal individuals, each ultimate summand, so to speak, is a cardinal individual. In a sum of cardinal collectives, the summands are either collectives or cardinal individuals, so that, given the sum, one cannot recover them as ultimate summands uniquely. To circumvent this situation, let us introduce the notion of an *attributive* relational collective, essentially as in OM for attributive relation-individuals.

In *D9* above the general notion of an n-adic relational collective among cardinal individuals was introduced. In similar fashion, let us now introduce the general notion of an n-adic relational collective among cardinal collectives. We let

D14. 'RelColl(Coll(Card))$_n$ x' abbreviate '(y)((Unit y . y P x) $\supset$ (Ez_1)...(Ez_n)(Coll(Card) z_1 . ... . Coll(Card) z_n . y Ord $z_1,\ldots,z_n$))'.

An entity x is thus an n-adic relational collective with cardinal collectives as relata just where every unit part of x is an ordinal entity with just n cardinal collectives as ordinands. An attributive relational collective of cardinal collectives is now a dyadic relational collective of cardinal collectives of a special sort.

The familiar way of writing dyadic-relation formulae is to allow the relation term to occur in the middle flanked by argument-terms on either side. To achieve this for relations between cardinal collectives, we may let

D15a. '$x\ y_2\ z$' abbreviate '(RelColl(Coll(Card))$_2$ y . Coll(Card) x . Coll(Card) z . (Ew)(w Ord x,z . w P y))'.

In a similar fashion, we may define '$x\ y_3\ z,w$', and so on, for triadic relational collectives among cardinal collectives, and so on.

We now define

D16. 'x_2 Att$_{\text{Coll(Card)}}$ y' as '($y\ x_2\ y$. (z)(w)($z\ x_2\ w \supset z = w$))',

so that x_2 is an attribute of the cardinal collective y just where x_2 is a RelColl(Card))$_2$ that the cardinal collective y bears to itself, and such that any two cardinal collectives one of which bears it to the other are in fact one and the same.

Given this special notion of attribute, we may introduce the relation *Suc of the *ancestral* of the relation Suc.[4] (We write 'Att' here as short for 'Att$_{\text{Coll(Card)}}$'.)

D17. 'x *Suc y' abbreviates '(Coll(Card) x . Coll(Card) y . (u)((u_2 Att y . (z)(w)((u_2 Att w . z Suc w) $\supset$ u_2 Att z)) $\supset$ u_2 Att x))'.

The definiendum expresses that the Coll(Card) x bears the ancestral of Suc to the Coll(Card) y, or equivalently, that x can be reached in a finite number of Suc-steps from y.

Readers familiar with the notion of the ancestral, and of the notion of the ancestral for relations of higher degree n where n is an even integer, will recognize how easily we can go on, from this point, to develop the arithmetic of the finite integers. (See VII below and the Appendix.)

Let us reflect now upon how a version of the simplified theory of types may be developed here within HL, merely by extending the definitions already at hand by giving further ones of similar kind. For this let us distinguish the *level* or *argument-structure* of the various kinds of collectives to be accomodated. Let 1 indicate the level of the cardinal individuals and collectives of such, and let (1,1) indicate the level of structure of a dyadic ordinal individual with cardinal ordinands or any collectives (sums) of such, including attributive relations. Similarly (1,1,1) is the level for triadic ordinal individuals with cardinal ordinands. Then ((1,1),(1,1)) indicates the level or argument-structure of dyadic relational collectives between dyadic relational collectives between cardinal individuals. And similarly for all the further kinds of *n*-adic relational collectives of all levels and kinds of argument-structures, including all manner of further attributive relations.

This is not the occasion to tarry with the development of the system into the foundations of mathematics. Such development may be carried out essentially as in EM, *mutatis mutandis*. Along the line a *Multiplicative Axiom* will be needed, as well as further notions and principles for extensions into transfinite number theory and into the theory of infinitesimals.

For purposes of application to non-mathematical material, suitable individual-constants may be added as primitives, and erstwhile predicate-constants added as primitive constants for the appropriate kinds of collectives. We thus have in HL a foundation for applied logic as extensive as that of *PM* itself.

The underlying logic in the formulation of HL here has been that of the usual first-order quantification-theory. It is easy to see, however, how this could be transformed into a relational algebra, as formulated previously.[5] In place of '~', 'v', and the quantifiers as primitives, we now need the Boolean notions '-' and '∪' for the negatives and sums of *n*-adic relations, the notion '×' for the Cartesian product of any *k*-adic relation with an *n*-adic one, and the general notion of the relations *affilated* with a given *n*-adic relation, i.e., all its domains, relational domains, and converses. The algebra of relations provides the needed logical background without the use of variables, truth-functional connectives, or quantifiers. The two non-logical primitives here are 'P' and 'Ord' and the whole of the theory of types now becomes a relational algebra rather than a logic or applied logic of the usual kind.[6]

CHAPTER SIX

On non-translational semantics and a quasi-Leśniewskian calculus of names

> "On se persuade mieux pour l'ordinaire par les raison qu'on a trouvé soi-meme, que par celles qui sont venues dans l'esprit des autres."

According to Lejewski[1], Leśniewski's so-called Ontology "has been described as the most comprehensive Logic [or Calculus] of Names because its most characteristic expressions of Protothetic belong to the semantical category of propositions." Protothetics, it will be recalled, is in essentials Leśniewski's rendition of the classical two-valued logic of the truth-functional connectives. As Lejewski points out, however, Protothetic "goes beyond the classical Calculus of Propositions in allowing variables over functions upon propositions as well as over propositions themselves, together with quantifiers over both of these kinds of entities." Just as propositions are the values for variables in the logic of propositions, so, it would seem, should names be the values for variables in Ontology regarded as a Logic of Names. The sense in which Ontology is to be regarded as a Logic or Calculus of Names, however, is far from clear.

Leśniewski's 'ε' for 'is' is the basic context '$x \varepsilon y$' is not, *prima facie* anyhow, to be read in terms of names. To say that $x \varepsilon y$ is not to say that the name x is the name y, but rather that x *is* y. By a considerable circumlocution we might "read" '$x \varepsilon y$' in terms of names, perhaps to the effect that some suitable name of x bears a certain semantical relation to a suitable name of y. Or, if 'x' and 'y' are taken as variables for names, then '$x \varepsilon y$' may be taken to express that an object of which x is a name *is* an object of which y is a name. For either "reading" here, a considerable spelling out of the theory is needed. If we mistake not, no such theory for the reading in terms of names has yet been explicitly given in modern terms by Leśniewski's commentators.

It is not sufficient, in formulating a philosophical or scientific theory, merely to lay down a primitive and say that it is to be "read" in context in such and such a way. No, it is essential explicitly to spell out the "interpretation" of that term in a well-formulated semantical metalanguage, which itself should be of a certain simplicity and acceptability. The standards for these latter may of course vary a good deal from philosopher to philosopher. But never mind: the more exacting our standards here, surely, the better. In any event, it behooves the Leśniewskian commentators to apprise us of the exact semantics needed for the "reading" of the basic contexts containing the primitive 'ε' and the "characteristic expressions" of the theory, and to assure us that these meet certain minimal desiderata. Not to do so is to resort to older, informal techniques in an area where sophisticated methodologies are now available.

This present paper is not concerned with the niceties of how to interpret Leśniewski. The attempt is made here, rather, to show that the quasi-Leśniewskian theory HL of the preceding chapter, can be formulated as a genuine calculus of names by bringing in the devices of non-translational semantics. This latter, it will be recalled, is a semantic lacking the crucial feature that all other kinds of semantics have, namely, of containing the object-language *in toto*, or a translation of it, as a part.[2] A non-translational semantics of the system HL will then turn out, as we shall see, to be a genuine calculus of names in a very direct sense. Its "characteristic expressions" include those for the vocabulary of HL, and the whole theory is a theory about those expressions. The theory, however, is not a mere syntax, although it includes such as a part. A syntax is of course a "calculus of names," but only in a more restricted sense.

There is no need here to go into laborious technical detail in the presentation. The formulation of HL presupposed is that of the preceding chapter, and the key items of non-translational semantics needed will be recalled one by one as we proceed.

The first item on the agenda is to be clear about the shape-descriptions to be admitted in the syntax. The expressions of HL include, it will be recalled, primitive constants for any or all the kinds of entities admitted, cardinal entities and ordinal ones of any complexity. The unit-sum terms include all expressions of the form '$(x1\text{--}x\text{--})$' where '--x--' is a sentential form of HL perhaps containing 'x' as a free variable. Shape-descriptions for all of these will play a central role in what follows.

To keep tabs on the structure of HL and the metalanguage to be formulated, let us recall the full syntactic character of the former by means of the following stipulation, by simultaneous recursion, of 'formula' and 'term' as relativized to that system.

1. Any variable or primitive individual-constant is a term of HL.
2. If **x** and **y** are terms, ⌜**x** P **y**⌝ is a formula of HL.
3. If **x**, **y**, and **z** are terms of HL, ⌜**x** Ord **y**,**z**⌝ is a formula of HL.
4. If **x** is a variable and ***A*** is formula of HL, then ⌜(**x**1***A***)⌝ is a term of HL.
5. If ***A*** is a formula of HL, so is ⌜~ ***A***⌝.
6. If ***A*** and ***B*** are formulae of HL, so is ⌜(***A*** v ***B***)⌝.
7. If **x** is a variable and ***A*** a formula of HL, then ⌜(**x**)***A***⌝ is a formula of HL.

The next item on the agenda is to give the full list of shape-descriptive names to be admitted in the syntax. These will include names for each of the primitive symbols of HL. These latter are just 'P', 'Ord', '1', '~', 'v', left and right parentheses, variables, and the primitive individual-constants. In particular, then, let us let 'PEE' be the shape-descriptive name for 'P'. 'ORD' for 'Ord', 'ONE' for '1', 'TILDE' for '~', 'VEE' for 'v', 'LP' for '(', 'RP' for ')', 'EX' for '*x*'. Let further variables be written as '*x*', '*x*′', and so on. We let 'AC' be the shape-descriptive name for '′'. Similarly let all primitive individual-constants of HL be written as 'a', 'a′', and so on. We then need 'AY' as the only additional structural-descriptive name for 'a'.

Let '*a*', '*b*', and so on, now be the syntactical variables for all the expressions of HL, and let '⌒', the printer's ligature sign, be the sign for concatenation. (*a*⌒*b*) is then the result of concatenating *a* with *b*. All concatenates of all the expressions of HL are assumed to be available as values for the syntactical variables. They are "available" in the sense of timelessness, so to speak, as "potentials" or potential entities, if one likes, but not necessarily actual ones exhibited anywhere in space or time or written down or directly experienced by anyone in any concrete way.

Let 'Sub' be the sole semantical primitive and significant in contexts of the form '*a* Sub *b*', expressing that the expression *a* is *subsumed* in the expression *b*. Within a semantics for HL based on the relation Des for designation[3], we could define '*a* Sub *b*' as

'(E*x*)(E*y*)(*a* Des *x* . *b* Des *y* . *x* P *y*)'.

Such a definition is not available here, however, the semantics being a non-translational one in which there is no talk about parts and wholes and no designation-relation. No, instead, we use 'Sub' as the sole mode of semantical locution over and above those in the underlying syntax. This latter, incidentally, is to include identity as a primitive. But we must then, of course, distinguish between '$a = b$', stating that a and b are the same expression, and '(a Sub b . b Sub a)', stating that a and b are co-extensive in the sense of being mutually subsumed in each other. Thus we may immediately let

'a CoExt b' abbreviate '(a Sub b . b Sub a)'.

This metalanguage may be referred to as SM^{HL}_{Sub}, the semantical metalanguage for HL based on Sub, its non-translationality being assured by the conditions to be laid down for Sub. The full character of SM^{HL}_{Sub} can be summarized in the following stipulation by simultaneous recursion of 'term' and 'formula', more particularly, of 'term of SM^{HL}_{Sub}' and 'formula of SM^{HL}_{Sub}'.

1. Any syntactical variable or shape-descriptive name of a primitive sign of HL is a term.
2. If $\boldsymbol{a}$ and $\boldsymbol{b}$ are terms, so is $\ulcorner(\boldsymbol{a}^\frown\boldsymbol{b})\urcorner$.
3. If $\boldsymbol{a}$ and $\boldsymbol{b}$ are terms, $\ulcorner\boldsymbol{a} = \boldsymbol{b}\urcorner$ and $\ulcorner\boldsymbol{a}$ Sub $\boldsymbol{b}\urcorner$ are formulae.
4. If $\boldsymbol{A}$ is a formula, so is $\ulcorner\sim \boldsymbol{A}\urcorner$.
5. If $\boldsymbol{A}$ and $\boldsymbol{B}$ are formulae, so is $\ulcorner(\boldsymbol{A} \vee \boldsymbol{B})\urcorner$.
6. If $\boldsymbol{A}$ is a formula and $\boldsymbol{a}$ a variable, $\ulcorner(\boldsymbol{a})\boldsymbol{A}\urcorner$ is a formula.

Concerning these various notions we may now assume the usual principles of classical syntax as based upon '$\frown$'.[4] And concerning 'Sub' we assume principles providing for a full Boolean algebra of names of a certain kind.

Let 'PrNm a' be defined now within the syntax to express that a is a primitive individual-constant (or name) of HL. And let 'UnitSumTrm a' be defined to express that a is a unit-sum term, i.e., an expression of HL of the form $\ulcorner(\boldsymbol{x}1\text{--}\boldsymbol{x}\text{--})\urcorner$ where $\boldsymbol{x}$ is a variable and $\ulcorner\text{--}\boldsymbol{x}\text{--}\urcorner$ a formula containing $\boldsymbol{x}$ as its only free variable if any. A unit-sum term containing no free variable is also a kind of constant, a *defined* name or constant, we might say. The only names we need recognize in HL are the primitive individual-constants and the unit-sum terms containing no free variables. Thus we may let

'Nm a' abbreviate '(PrNm a v UnitSumTrm a)'.

As a basic *Limitation Rule* concerning 'Sub', we then immediately assume that

⊢(*a*)(*b*)(*a* Sub *b* ⊃ (Nm *a* . Nm *b*)).

To see how a Boolean algebra of names may be supplied, let us note that we may let

'UnivNm *a*' abbreviate '(*b*)(Nm *b* ⊃ *b* Sub *a*)',
'NullNm *a*' abbreviate '(*b*)(Nm *b* ⊃ *a* Sub *b*)',
'*a* Sum *b*,*c*' abbreviate '(*b* Sub *a* . *c* Sub *a* . (*d*)((*b* Sub *d* . *c* Sub *d*) ⊃ *a* Sub *d*))',
'*a* Prod *b*,*c*' abbreviate '*a* Sub *b* . *a* Sub *c* . (*d*)((*d* Sub *b* . *d* Sub *c*) ⊃ *d* Sub *a*))',
'*a* Neg *b*' abbreviate '(*c*)(((*c* Sub *a* . *c* Sub *b*) ≡ NullNm *c*) . ((*a* Sub *c* . *b* Sub *c*) ≡ UnivNm *c*))'.

It is not difficult to see that these definitions give us Boolean analogues of the semantical notions of being a *universal* name, a *null* name, the semantical *sum* of two names, the semantical *product* of two names, and the semantical *negative* of a name. Suitable semantical principles concerning Sub may be laid down to assure that these defined notions have the Boolean properties required of them. The various axioms and rules needed for the underlying syntax are of course presupposed.

Let us note now how some of the key notions of HL may be mirrored here in the non-translational semantics. First, a *unit* name is as follows.

'UnitNm *a*' for '(Nm *a* . ~ NullNm a . (*b*)((Nm *b* . ~ NullNm *b* . *b* Sub *a*) ⊃ *a* Sub *b*))'.
'ORDENT' for '(LP⌢EX⌢ONE⌢TILDE⌢LP⌢EX⌢AC⌢RP⌢LP⌢EX⌢AC⌢AC⌢RP⌢LP⌢EX⌢AC⌢AC⌢AC⌢RP⌢LP⌢TILDE⌢EX⌢PEE⌢EX⌢AC⌢VEE⌢TILDE⌢EX⌢AC⌢ORD⌢EX⌢AC⌢AC⌢EX⌢AC⌢AC⌢AC⌢RP⌢RP⌢)' [i.e '*x*1 ~ (*x*′) (*x*″) (*x*″)(~ *x* P *x*′ v ~ *x*′ Ord *x*′ ,*x*‴))']'],[5]
'OrdEntNm *a*' for '*a* CoExt ORDENT',
'CardEntNm *a*' for '(*b*)((UnitNm *b* . *b* Sub *a*) ⊃ ~ OrdEntNm *b*)',
'CardIndNm *a*' for '(UnitNm *a* . CardEntNm *a*)',
'Coll(Card)Nm *a*' for '(*b*)((UnitNm *b* . *b* Sub *a*) ⊃ CardIndNm *b*)'.
'LesColl(Card)Nm *a*' for '(Coll(Card)Nm *a* . ~ NullNm *a* . ~ CardIndNm *a*)',
'FictNm *a*' for 'NullNm *a*'.

Clearly the definienda here express that *a* is a name for the intended kind of entity.

We have now the wherewithal with which to accommodate adaptations of the Leśniewskian "functors" of "inclusion", of "exclusion", of "identity", and of "existence". The functors of inclusion in the calculus of names are now as follows (*mutatis mutandis*, as in the preceding chapter.)

'*a* ε *b*' abbreviates '((CardIndNm *a* . CardIndNm *b* . *a* CoExt *b*) v (CardIndNm *a* . LesColl(Card)Nm *b* . *a* Sub *b*))',
'*a* ⊏ *b*' abbreviates '(*a* ε *b* v (LesColl(Card)Nm *a* . LesColl(Card)Nm *b* . (*a* CoExt *b* v *a* Sub *b*)))',
'*a* ⊂ *b*' abbreviates '(*a* ⊏ *b* v (((FictNm *a* . CardIndNm *b*) v (FictNm *a* . LesColl(Card)Nm *b*) v (FictNm *a* . FictNm *b*)) . ~ *a* CoExt *b*))',
'*a* △ *b*' abbreviates (*a* ⊏ *b* v '((LesColl(Card)Nm *a* . CardIndNm *b* . *b* Sub *a*) v (LesColl(Card)Nm *a* . LesColl(Card)Nm *b* . *b* Sub *a*) v (LesColl(Card)Nm *a* . LesColl(Card)Nm *b* . (E*c*)(LesColl(Card)Nm *c* . *c* Sub *a* . *c* Sub *b*)))'.

In similar fashion the three "functors of identity" are provided here as follows.

'$a =_{\text{Les}} b$' abbreviates '(CardIndNm *a* . CardIndNm *b* . *a* CoExt *b*)'
'*a* □ *b*' abbreviates '($a =_{\text{Les}} b$ v (LesColl(Card)Nm *a* . LesColl(Card)Nm *b* . *a* CoExt *b*))',
'*a* 0 *b*' abbreviates '(*a* □ *b* v (FictNm *a* . Fict Nm *b* . *a* CoExt *b*)).

And the three "functors of existence" are then:

'Ex *a*' for '(CardIndNm *a* v LesColl(Card)Nm *a*)',
'Sol *a*' for '(CardIndNm *a* v FictNm *a*)',
'Ob *a*' for 'CardIndNm *a*'.

With these definitions our play is about done, in the sense that a calculus of names of a quasi-Leśniewskian kind is seen to be incorporated within a non-translational semantics for HL. Suitable rules governing Sub may of course be laid down and the various Leśniewskian Axioms of Ontology then seen to be consequences of them.

We should observe, however, that a non-translational truth-predicate is immediately forthcoming in terms of 'Sub'. In fact we can let

'Tr a' abbreviate '(Sent a . (Eb)(Ec)(UnivNm b . Vbl c . b Sub (c⌢ONE⌢a))'.

'Sent a' here of course expresses that a is a sentence of HL, and 'Vbl c' that c is a variable. Thus a is true if and only if some universal name is subsumed in some UnitSumTrm (c⌢ONE⌢a), where c is a variable.

The desired properties of Tr are now forthcoming, but of course no Adequacy Condition, of the kind occurring in a translational semantics — a generalized 'Tr 'snow is white' ≡ snow is white' principle — is available here. Such a condition cannot even be stated, for it requires both the use and mention of the object-language expressions. In a non-translational semantics, there is no use made of the object-language expressions, the theory being solely one concerning their mention, so to speak.

In the detailed development of the non-translational semantics of the unit-sum terms, however, we should along the way meet with some principles that would bring us close to an Adequacy Principle. Thus, at some point we would meet with principles to the effect that

⊢ (a)(b)(c)((Sent a . UnivNm b . Vbl c) ⊃ (b Sub (c⌢ONE⌢a) v b Sub (c⌢ONE⌢TILDE⌢a))),

and that

⊢ (a)(b)(c)((Sent a . UnivNm b . Vbl c) ⊃ (b Sub (c⌢ONE⌢a) ≡ ~ b Sub (c⌢ONE⌢TILDE⌢a))).

These are in effect (non-translational) semantical forms of laws of excluded middle and of non-contradiction. But also we would meet with more special principles, such as that

⊢ (a)(b)(c)(d)((Nm a . Nm b . UnivNm c . Vbl d) ⊃ (a Sub b ≡ d Sub (d⌢ONE⌢a⌢PEE⌢b))),

from which we could easily prove that

⊢ (a)(b)((Nm a . Nm b) ⊃ (a Sub b ≡ Tr (a⌢Pee⌢b))).

And similarly for disjunctions, that

⊢ (a)(b)(c)(d)((Nm a . Nm b . Nm c . Nm d) ⊃ ((a Sub b v c Sub d) ≡ Tr (LP⌢a⌢PEE⌢b⌢VEE⌢c⌢PEE⌢d⌢RP))),

and for negations, that

⊢(a)(b)((Nm a . Nm (b) ⊃ (~ a Sub (b ≡ Tr (TILDE⌢a⌢PEE⌢b))).

And similarly for negations of disjunctions, disjunctions of negations, and so on. Although these principles are not strictly Adequacy Principles, they do show how the use of 'Sub' in the metalanguage can in a certain way take the place of 'P' in the object-language. They are special principles of adequacy for object-language sentences containing only atomic and molecular formulae built out of 'P' without using 'Ord'.

For sentences containing quantifiers, we have that

> $\vdash(a)(b)((\mathrm{Vbl}\ a\ .\ \mathrm{Nm}\ b) \supset (\mathrm{Tr}\ (\mathrm{LP}{}^\frown a{}^\frown \mathrm{RP}{}^\frown a{}^\frown \mathrm{PEE}{}^\frown b) \equiv (c)(\mathrm{Nm}\ c \supset c\ \mathrm{Sub}\ b))),$

and, more generally, that

> $\vdash\ulcorner(a_1)\ldots(a_n)((\mathrm{Vbl}\ a_1\ .\ \ldots\ .\ \mathrm{Vbl}\ a_n) \supset(\mathrm{Tr}\ (\mathrm{LP}{}^\frown a_1{}^\frown \mathrm{RP}{}^\frown \ldots{}^\frown \mathrm{LP}{}^\frown a_n{}^\frown \mathrm{RP}{}^\frown \boldsymbol{B}) \equiv (c_1)..(c_n)((\mathrm{Nm}\ c_1\ .\ \ldots\ .\ \mathrm{Nm}\,c_n) \supset \text{--}c_1\text{--}\ldots\text{--}c_n\text{--})))\urcorner$, where '$\text{--}c_1\text{--}\ldots\text{--}c_n\text{--}$' is a formula of $\mathrm{SM}^{\mathrm{HL}}_{\mathrm{Sub}}$ containing only '~', 'v', 'Sub', and the variables 'c_1', ..., 'c_n', and where in place of '***B***' a shape-description is inserted of a formula of HL containing 'P' and HL-variables wherever there are occurrences of 'Sub' and of 'c_1', ..., 'c_n' in '$\text{--}c_1\text{--}\ldots\text{--}c_n\text{--}$'.

The proviso here requires of course that the proper shape-description corresponds appropriately with the given formula of $\mathrm{SM}^{\mathrm{HL}}_{\mathrm{Sub}}$. These principles are in effect principles concerning the so-called "substitutional interpretation" of the quantifiers, about which there has been a good deal of sound and fury in recent years without much clear-cut semantics for it by way of underlying theory.

It should be noted that in the foregoing no use is made of Leśniewski's Protothetic, with its additional variables and quantifiers. The use of variables ranging over some mysterious realm of entities called 'propositions' is thought to be theoretically unsound. Whatever propositions are, they are a semantical kind of entity, to be accommodated — if at all — as suitable constructs in a semantical or pragmatical metalanguage.[6] For Leśniewski, it seems, propositions are needed right at the beginning, and are in effect postulated as unanalyzed entities *sui generis*. Further, in Protothetic variables are admitted also, as already noted, over functions of propositions, i.e., over properties of and relations between or among them — not only truth-functional ones, but presumably all manner of intensional and pragmatical ones as well. There is a too-muchness about this sort of a beginning, utilizing obscure items that do not become clarified even in the full peroration of the system.

Also we should note that all the "semantical categories" needed in the system here are declared *ab initio*, and not *seriatim* in the development of it. As Stachniak has remarked, in a recent monograph, "the introduction of a new category to the system [of Ontology] causes every time the change of language alphabet, rules of inference and the sets of well-formed expressions and formulas. This makes semantical investigations of the system more difficult."[7] Indeed it does, for with a change of vocabulary (and hence of the formulae) and of rules of inference, one has a change of system — and strictly we are no longer dealing with the same system at all. The introduction of a new semantical category requires a new system, an extension (in some sense) of the preceding one perhaps, but new nonetheless. By contrast, HL and SM^{HL}_{Sub} are fixed systems with all their "semantical categories" intact. Of course HL may be applied to new non-logical subject-matters — to physics, to angelology, or whatever — but this is another matter altogether.

Another important point of contrast is of course in the handling of relations. The ordinal individuals supply the wherewithal within HL of building up a very general theory of relations, whereas the system of Ontology proper contains no relations at all until some new "semantical category" of them is declared with the attendant change of system required to handle them.

It should be emphasized again that, although this paper revolves around the name of Leśniewski, the systems HL and SM^{HL}_{Sub} are not, in any strict sense, intended to capture his views. They merely approximate them to some extent and in some respects, and bear some family resemblance to them. Even so, the kindship between a calculus of names and non-translational semantics seems so striking that it is perhaps surprising that no Leśniewski-commentator has pointed it out heretofore.

CHAPTER SEVEN

On platonism, form, and number

"Periculosae plenum opus aleae Tractas, et incedis per ignes Suppositos cineri doloso."

Platonism, or platonic realism, in logic and mathematics has been closely associated with the names of Frege, Zermelo, Russell, Carnap, Gödel, Bernays, and Quine. Mathematical platonism has popularly become identified with the acceptance of an ontology of sets and/or classes and/or relations as fundamental constituents of "the furniture of the world", to use Russell's famous phrase. The phrase is not a happy one, however, furniture being provided for — usually anyhow — only *after* the edifice has been constructed. Genuine platonism should, it might be thought, take sets and/or classes and/or relations as fundamental among the building materials of the cosmos and all that is therein. Usually, also, these entities are regarded as "abstract" rather than "concrete," but no one seems to have given us a detailed account as to just how this dichotomy is to be characterized. Sets, classes, and relations somehow are taken as the paradigm of "abstract" entities, as over and against something called 'individuals', which are then the paradigm for "concrete" ones. This is all very unsatisfactory and in need of much clarification.

Another facet of platonism in mathematics was called attention to by Paul Bernays in his famous paper "On Platonism in Mathematics," back in 1934, namely that of "viewing the objects [of mathematics] as cut off from all links with the reflecting subject."[1] Strictly, however, mathematics, no matter how abstractly conceived, is never wholly cut off in this way. There is always a mathematician floating around somewhere doing the mathematics, dreaming it up, proving theorems, applying them, and, above all, criticizing the mathematics done by others as either "false, trivial, or mentioned in a footnote in one of my [his or her] papers of several years back!"

Bernay's point is, of course, that the mathematician is not mentioned by name and is not a value for a parameter occurring in the actual formulae. Pure mathematics in the classical sense is independent of the "reflecting subject" in this way, surely, whereas mathematical intuitionism, in some formulations anyhow, is not.[2] Curiously, however, the formulations of intuitionism that are not, i.e., those that do contain formulae with parameters for the "creative subject" or subjects whose mathematical "constructions" are under discussion, came on the market only years after Bernays wrote. The formulations of intuitionism prior to 1934 lacked this feature, however, and hence should be classified as platonistic according to Bernays' criterion — contrary presumably to his intent.

If we are far from being clear as to how "abstract" objects differ from "concrete" ones and if mathematics — at least applied mathematics — is never wholly cut off from all links with the "creative" or "reflective" subject, what then is mathematical platonism? it may well be asked. Worse yet, sets and/or classes do not *sans phrase* appear to be prototypes of Plato's forms or *idéa* or *eidé*, although they are commonly taken to be. A considerable argument and confrontation with the actual text is surely needed to establish that they are, if they are. Meanwhile we may be on rather shaky grounds, it would seem, if we assume that they are. Note that we have not spoken of *relations* in this context. Relations seem to be eminently platonic entities, it would seem, in a sense in which sets and/or classes are not. Goodness and justice are not to be construed as sets or classes yet are intimately interrelated with each other. The numbers 1 and 2 need not be regarded a sets, yet arithmetic would be impossible were they not taken to stand to one another in certain relations.

It is often said that Plato had no theory of relations in the modern sense. He made sophisticated use of them, however, in the sense of naming some of them and even of being aware of some of their key logical properties, such as symmetry or asymmetry. Be this as it may, it is the theory of relations that *par exellence* distinguishes modern logic from its ancient and medieval counterpart of syllogistics. Graft a theory of relations on to syllogistics in one way or another, and a respectable modern logic may emerge.[3] Curiously enough, even up to the very moment the general theory of relations is rather neglected in most histories of logic and in almost all textbooks, yet, as Russell noted in *Mysticism and Logic* (1929, pp. 76-77), it was "invented [or formulated] to deal with topics that wholly surpassed the powers of the old logic, though they [these topics] form the chief contents of mathematics." Plato's *sotto voce* recognition of relations and his concern with mathematics are no doubt intimately intertwined.

According to A.E. Taylor[4], the text of the *Epinomis* 990c5-991b4, is "the most important mathematical passage in the platonic *corpus*," whether actually written by Plato or not. However, the text is "unfortunately uncertain, in part probably corrupted, in part also probably never reduced to grammatical form by the writer." Even so, Taylor comments, "the sense is clear," although it is difficult to determine just what the sense is. As Taylor himself urges (p. 516), "it is time that we should make an end of the pretence of understanding [Plato] by ignoring the [textual] evidence or by arbitrarily reading into ... [it] the views of our own favorite modern metaphysicians." At the same time we cannot hope to understand the mathematical portions of the Platonic *corpus* without taking into accont the full, broad historical context, including our modern knowledge of the subject. Without taking what we now know into account, we may be doing Plato grave historical injustice, by not recognizing the character of his "anticipations" and the depth of some of his basic insights. To anticipate modern developments with depth and acumen is often as much a mark of historical greatness as full-bodied achievement is. (Peirce is in many respects as great as Frege, the former being a great anticipator, the latter a great achiever.) In any case, let us follow A.E. Taylor for the moment — an admirable Virgilian guide, surely, in spirit if not always in letter — in his general reading of the Platonic *corpus*, including the *Epinomis*, as being in principle harmonizable with the modern theory of number, stemming — in broad outline anyhow — from the work of Frege, Dedekind, and Whitehead-and-Russell, in which the modern logic of relations plays a fundamental role.

Two items here are of especial importance. First, we must seek some clarity concerning the character and inner structure of the integers themselves. Then, secondly, we must understand the importance of Plato's "revolutionary demand [fn., p.501] that quadratic and cubic surds [e.g., $\sqrt{2}$, $\sqrt[3]{3}$] shall be recognized [also] as numbers in opposition to the traditional view that there are "irrational" magnitudes (length, areas, volumes), but no "irrational" numbers."

Let us reflect upon the first first. According to Aristotle (*Metaph.* M 1083a32), Plato is said to have said that "there is a first 2 and first 3, and the numbers are not addible [as units] to one another," And, somewhat similarly in the *Ethica Nicomachia* (1096*a*17 ff.), we are told that there is no form (*idéa*) of number, because "in numbers there is a before and an after" (a priority and posteriority). Taylor interprets these contentions as amounting to the doctrine that numbers form a *series*. More correctly, we should say here that the relation of being-less-than as between numbers is a serial relation, i.e., a connected, asymmetrical, transitive dyadic relation. The sense of these contentions, according to Taylor, is then that "the series of numbers is not made by adding "units" together. *E.g.* we say that 3 + 1 = 4, but we do not mean that 3 is three "units" or that 4 is 3 *and* 1 [where the 'and' is the 'and' of *withness* or of being *accompanied by*]; 4 is not four 1's, or a 3 and a 1; it is *one* 4. What we really *add* together is not numbers but aggregates or collections," he goes on. "Thus it is true that if you have a group of *n* things and another group of *m* things, and form the two into one group, this new group contains *m* + *n* things, but it is not true that the *number m* + *n* contains a number *m* and a number *n* [as its parts. No, we should say rather that the numerical expression '*m* + *n*', where '*m* and '*n*' are constants, has '*m*' and '*n*' as its summands.] The importance of this view is that it leads to revision of the whole conception of number. The fifth-century theory, still represented by Euclid's definition of *arithmos* (Elements vii, def. 2), is that a "number" is *plethos monadon*, a "collection of 1's". On the new view," which Taylor seems to think present in Plato if only implicitly, "no number is a "collection"; the statement that 3 = 2 + 1, which is the definition of 3, does not mean that 3 is "a 2 and a 1," but that 3 is the term in the integer-series which comes "next after" 2." The introduction of the relation of coming "next after" here is a very bold step and leads us at once to the portals of modern formulations of arithmetic, as we shall see in a moment.

Let us now turn to Taylor's second contention concerning surds and irrational magnitudes, as discussed especially in the *Philebus* and the *Epinomis*. Taylor sees in the Dedekindian notion of a "section," or "cut," notions exhibiting the character Plato ascribes to, say, $\sqrt{2}$. "We cannot find a rational fraction the "square" of which is exactly 2," Taylor reminds us (p. 511) in modern terms. "But we *can* divide all [positive] rational fractions into two classes, those of which the "squares" are less than 2 and those of which the squares are not less than 2. We see at once that the first of these sets has no highest term, the second no lowest, and that no fraction can belong either to both sets or to neither; thus our "section" is unambiguous, *i.e.* every fraction falls into one and only one of the two sets thus constituted. We may then define the "square root of 2" either as this section itself, or, if we prefer it, as the set of fractions whose squares are less (or, if we like, greater) then 2. ... [The "section"] involves an "indeterminate duality" (*aoristos dyas*) or "great-and-small," [namely,] the two sets, one of which has all its terms less than, the other all greater than, a specified value, and the duality is "indefinite" (*apeiron*) because one of the sets has no highest term, the other no lowest. The section is a determination of the "great-and-small" of the fractions by the "one" precisely because it makes an unambiguous "cut" just where it does. Other cuts can be made at other places in the series [of rationals], and each will define a different "real number.""

Note the subtle way in which 'set' and 'class' have crept into Taylor's description of the modern view, but assiduously avoided in favor of the more innocent 'aggregate', 'collection', and 'group' in his direct discussion of Plato. This is not accidental, it would seem. Aggregates or collections or groups of things are somehow things *summed together non-numerically*. This may be done without reifying the result into some kind of entity of a fundamentally new logico-metaphysical kind, a set or class. To do this latter seems somewhat contrary to the general tenor of Greek thought. The notion of set or class as such seems to be a primarily Germanic invention, a *Menge* being some object of a kind altogether different than that of its "members." Taylor can thus slip easily into talk of sets and classes in speaking of Dedekindian cuts, which are, after all, a product of German mathematical inventiveness. If he were to have availed himself of the English approach, especially that of his great compatriot George Boole, Taylor could have made use of a notion of non-numerical summation, Boole being apparently the first person in all history to have done so in an approximatively acceptable way.

Taylor's contentions here constitute, it would seem, a brilliant piece of exegetical reconstruction. Can they be substantiated on the basis of the actual text? or upon the basis of the teachings of the Later Academy? These are very delicate questions. Perhaps the best we can say is that they seem to accord with them and hence are not an "arbitrary" reading into the text of some preferred modern logico-metaphysical view. No, the reconstruction is one within pure mathematics itself, but can be carried out in a number of different ways technically. The problem of harmonizing one or the other of these ways with one's favorite modern view then remains as one requiring its own inner technicalia.

Let us go on now, in the present paper, to construct a Platonic theory of forms without sets or classes, in which, however, relations between or among forms play a fundamental role. This will be done in such a way as to take forms and relations together as entities of the same logico-metaphysical type or order, but without introducing *ad hoc* devices to avoid the Russell and other paradoxes. The formal details of the theory are essentially those of the systems OM and HL previously formulated.[5] The theory here in fact results from a reinterpretation of those systems in which the values for the variables are taken as forms or *eidé*, including relational ones, rather than as individuals (including relational individuals) and their parts as in OM, or as just individuals (including relational ones) as in HL. Further, we shall see how a theory of number emerges from the theory of forms, numbers themselves being of the same logico-metaphysical type. Numbers are thus seen to constitute a special subdivision or species of forms without invoking from without an artificial hierarchy for them. Such hierarchical subdivisions as are needed emerge from *within* the theory, so to speak, and depend upon the very natural way in which relational forms are provided for. All numbers emerge as forms, but not conversely.

The main novelty here is to provide for relation in terms of relational forms. Suppose *F* and *G* are forms in the usual sense, i.e., monadic ones. Another type of form is immediately forthcoming by combining *F* and *G* into an *ordinal* form having *F* and *G* as the ordinands in that order, or, conversely, *G* and *F*. Let '*H* Ord *F,G*' express that *H* is the form consisting of *F* and *G* taken in that order. 'Ord' will figure as a primitive in the system along with a general 'Incld' for *being included in*. '*F* Incld *G*' will express that the form *F* is included in the form *G*. *Equality* of forms is then merely mutual inclusion. In terms of inclusion, it is easy to show, a full Boolean algebra of forms may be developed.

The introduction of ordinal forms seems a very natural step to take in the development of a general theory of forms. Forms are already presumably "abstract" in some sense, so that whatever additional "abstraction" might reside in an ordinal form should cause no alarm. The ordinal form of *F* with *G* is closely akin to the ordered couple of them, i.e., the unit relation that *F* bears to *G*. (We can also speak here of the *virtual* ordered couple of or unit relation between them.[6]) The use of the relation Ord will enable us to gain the effect of having ordered couples, etc., without introducing entities of a type higher than that of *F*, *G*, and so on. By iteration, of course, we can go on to accommodate ordinal triples, quadruples, and so on.

Although the technical development here follows closely that of OM and HL, the differences in detail, in interpretation, and in terminology are sufficient to justify presenting the system *ab initio* for the benefit of readers not familiar with those systems.

As variables for forms let us use '*F*', '*G*', '*H*', '*R*', '*S*', and '*T*', with or without primes or numerical subscripts, taken as ranging over forms including relational ones. As the underlying logic let us presuppose the usual first-order quantification-theory. In addition, let us presuppose an operation of *unit summation* so that '(*F*1--*F*--)' is a term for all *unit* forms *F* such that --*F*--, where '--*F*--' is a sentential form or function of the system containing '*F*' as a free variable (if it contains any free variables at all). Just what a unit form is will be discussed in a moment.

Clearly now we can let

D1. '*F* Eq *G*' abbreviate '(*F* Incld *G* . *G* Incld *F*)',

so that equality is merely mutual inclusion, as already remarked.

The various Boolean operations on forms are immediately forthcoming in terms of unit summation.

D2a. '($F \cup G$)' abbreviates '(*H*1(*H* Incld *F* v *H* Incld *G*))',
D2b. '($F \cap G$)' abbreviates '(*H*1(*H* Incld *F* . *H* Incld *G*))',
D2c. '-*F*' abbreviates '(*G*1 ~ *G* Incld *F*)',
D2d. 'NFrm' abbreviates '(*F*1 ~ *F* Incld *F*)',

and

D2e. 'UFrm' abbreviates '-NFrm'.

These definitions introduce, respectively, the Boolean sum and product of forms *F* and *G*, the Boolean negation of form *F*, and null and universal forms.

What now is a unit form? A form in which one and only one "individual" participates, we might answer, taking 'individual' as the blanket word for whatever there is in the Platonic cosmos that is not a form — sensibles and other denizens of the lower realm — and taking 'participates' as naming the relation between individuals and the forms in which they inhere. For the moment we are considering only the theory of forms by itself, but later will add a new primitive 'Prtcp' for participation and a new domain of objects as "individuals." A unit form may be explained, however, quite independently of all talk of participation and individuals, as follows. We may let, namely,

D3. 'UnitFrm *F*' abbreviate '(~ *F* Incld NFrm . (*G*)((~ *G* Incld NFrm . *G* Incld *F*) ⊃ *F* Incld *G*))'.

This definition is of course closely analogous to the corresponding definitions in OM and HL.[7]

Let us now distinguish between *cardinal* and *ordinal* forms as follows.

D4. 'OrdFrm *H*' abbreviates '(E*F*)(E*G*) *H* Ord *F*,*G*'

and

D5. 'CardFrm *H*' abbreviates '(*G*)((UnitFrm *G* . *G* Incld *H*) ⊃ ~ OrdFrm *G*)'.

An ordinal form *H* is thus a form that bears Ord to any two forms *F* and *G* including, of course, the cases where *F* or *G* or both are themselves ordinal forms. A cardinal form, on the one hand, is a form such that all unit forms included in it are not ordinal forms. The aptness of this terminology is surely apparent.

The basic principles of the system are direct adaptations of those of OM and HL, and need not be given here, in spite of significant differences in meaning.

Much of this material may seem rather remote from Plato. Any reconstruction of the number-system in modern terms in the general manner of Frege or Whitehead-Russell, however, presupposes an extensive logical theory of some kind by way of background. Taylor followed these illustrious writers in making full use of sets and/or classes, or rather of more fundamental intensional entities of some kind or other, or the like, in terms of which the numbers may be introduced definitionally. Here cardinal and ordinal forms, together with non-numerical Boolean sums of such, are paid overtime, so to speak, to play the erstwhile role of sets or classes. The use of such sums seems closer — does it not? — to the spirit of Plato and of Greek mathematics generally than do the Taylorian sets and/or classes.

Let us turn now to the notion of a cardinal-unit *collective*, regarded as the NFrm or a Boolean sum of cardinal-unit forms. For this we may let

D6. 'Coll(CardUnit) *F*' abbreviate '(*G*)((UnitFrm *G* . *G* Incld *F*) ⊃ CardFrm *G*)'.

Similarly there are various kinds of *relational collectives*, depending upon the degree of the unit relations involved. To define formally the general notion of a relational collective of degree *n*, we first let

D7a. 'F Ord G,H,H'' abbreviate '(EF')(F Ord G,F' . F' Ord H,H')',

D7b. 'F Ord G,H,G',H'' abbreviate '(EF')(F Ord G,H,F' . F' Ord G',H')',

and so on. Then

D8. 'OrdFrm(CardUnit)$_n$ F' may abbreviate '(EG_1)...(EG_n)(F Ord $G_1,\ldots,G_n$. CardFrm G_1 . UnitFrm G_1 . ... . CardFrm G_n . UnitFrm G_n)'.

F is thus an n-adic ordinal form with cardinal-unit forms as its ordinands just where F bears Ord to n cardinal-unit forms, so to speak. A relational collective of degree n whose ordinands are cardinal-unit forms may now be introduced by letting

D9. 'RelColl(Card(Unit)$_n$ F' abbreviate '(G)((UnitFrm G . G Incld F) $\supset$ OrdFrm(CardUnit)$_n$ G)'.

Next let us introduce the notion of *being a successor of* as between cardinal forms. For this, we may let immediately

D10. 'F Suc$_{\text{CardFrm}}$ G' abbreviate '(CardFrm F . CardFrm G . (EH)(CardFrm H . UnitFrm H . ~ H Incld G . $F = (G \cup H)$)'.

This is not, of course, the successor relation of arithmetic. A cardinal form has many successors, whereas a natural number has one and only one successor in its series — cf. *D20* below.

We should note now the need for two Existence Principles, as follows.

$\vdash$(EF)(CardFrm F . UnitFrm F),

and

$\vdash$(F)(CardFrm $F \supset$ (EG)G Suc$_{\text{CardFrm}}$ F).

This latter will be recognized as a kind of *Axiom of Infinity*, assuring that every cardinal form has a successor. And the former, an unconditional *Existence Principle*, assures the existence of at least one cardinal-unit form, and hence of at least one cardinal form and of at least one unit form.

We now need to provide in general for the familiar notation involving relations, in which the relation-term occurs after the first term but followed by the remaining argument-terms on its right. This may easily be achieved for dyadic relations between cardinal collectives by letting

D11a. 'F H_2 G' abbreviate '(RelColl(CardUnit)$_2$ H . Coll(Card-

Unit) F . Coll(CardUnit) G . (EH')(H' Ord F,G . H' Incld H))'.

And similarly for triadic relations among cardinal-unit collectives, so that

D11b. '$F\ H_3G,F'$' abbreviates '(Relcoll(CardUnit)$_3$ H . Coll(CardUnit) F . Coll(CardUnit) G . Coll(CardUnit) F' . (EH')(H' Ord F,G,F' . H' Incld H))'.

And so on for relations of higher degree.

A significant technical point about quantification is to be noted as follows. Quantification over ordinal and relational forms of cardinal-unit collectives — these latter being merely Boolean sums of the former — is immediately forthcoming merely by quantifying the 'F' in contexts containing it in the context 'F_2'. Thus

D11a. '(F_2)--F--' is in effect short for '(F)(RelColl(CardUnit) F_2 ⊃ --F--)', where the only free occurrences of 'F' in '--F--' are in the context 'F_2',

and

D11b. '(EF_2)--F_2--' for '(EF)(RelColl(CardUnit)$_2$$F$. --F--)', where (etc.).

And similarly for triadic relations, and so on.

Suppose, however, we wish to quantify over a *property* or *attribute* of a cardinal-unit collective, and not just over a relation between or among such. Well, such properties here may be construed merely as *attributive* relations. Let

D12. 'F_2 AttRelFrm(CardUnit) G' abbreviate '($G\ F_2G$. (H)(H')($H\ F_2\ H'$ ⊃ H Eq H'))'.

Then F_2 is an attributive-relational form applying to G just where G bears F_2 to G itself, and in general where any two forms are equal that bear F_2 to each other.

It is very important to distinguish here between saying that a given form F is included in some form G and that F is one of the forms to which such and such an attribute is applicable. If F is a unit then the two are the same: F is included in G if and only if F is one of the unit forms comprising G or to which G is applicable. If F is not a unit, however, the two are not the same: F may be included in G without G's being attributable to F.

Let us think of individuals for the moment as cardinal-unit forms. The persons Plato and Aristotle, say, thus become identified with the cardinal-unit forms uniquely determinative of them. The compound form (Plato ∪ Aristotle) then is included in the form Greek, but this latter is not attributable to the former. (Plato ∪ Aristotle) is not a Greek, even though Plato and Aristotle separately are. However, if *F* is a dyadic *ordinal* form with Plato as its two ordinands, then to say that *F* is included in the form Greek (taking this latter now as an attributive-relational form in the sense of *D12*) is in fact to say that the form Greek is attributable to Plato. In effect 'Greek' here can be read 'is the same Greek as'. Thus it holds that

(Greek AttRelColl(CardUnit) Plato ≡ (E*F*)(*F* Ord Plato,Plato . *F* Incld Greek)).

Quantifiers over attributes can now be handled merely as special cases of *D11a* and *D11b*.

We are now in a position to turn to the cardinal numbers, in the sense of the numbers *of* given collectives, so that, as with Taylor, if we "have a group of *n* things and have another group of *m* things, this new group contains *m* + *n* things." What now is a group or collective of *n* things, according to the present theory? Well, in terms of the adaptation of Frege and Whitehead-and-Russell here, we can say that it is a collective of cardinal-unit forms one-to-one correspondible with some given collective of *n* cardinal-unit forms. How now can this notion be introduced?

We need the notion of a one-to-one correspondence or correlation between unit-cardinal collectives. For this, in more or less standard fashion, we may let

D12. 'One-Many F_2' abbreviate '$(G)(H)(G')((G\ F_2H\ .\ G'\ F_2\ H) \supset G \text{ Eq } G')$',

D13. 'Many-One F_2' abbreviate '$(G)(H)(H')((G\ F_2\ H\ .\ G\ F_2\ H') \supset H \text{ Eq } H')$',

and

D14. 'One-One F_2' abbreviate '(One-Many F_2 . Many-One F_2)'.

A one-to-one relation F_2 is thus both a one-to-many correspondence as well as a many-to-one one.

What now do we mean by the *equinumerosity* (or *similarity* in essentially the Whitehead-Russell sense, *PM* *73.01) of two cardinal-unit collectives *F* and *G*? Well, again in the adaptation here, just where there is a suitable one-to-one correlation between their respective cardinal-unit components. Thus we may let

D15. 'F Equinum G' abbreviate '(Coll(CardUnit) F . Coll(CardUnit) G . (EF')(RelColl(CardUnit)$_2$ F' . One-One F_2' . (H)(H')(H F_2' H' ⊃ (UnitFrm H . CardFrm H . H Incld F . UnitFrm H' . CardFrm H' . H' Incld G)) . (H)((UnitFrm H . H Incld F) ⊃ (EH')(UnitFrm H' . H' Incld G . H F_2' H')) . (H')((UnitFrm H' . H' Incld G) ⊃ (EH)(UnitFrm H . H Incld F . H F_2' H')))'.

Finally now, we may ask, what is the *cardinal number of* a given collective F of cardinal-unit forms? Well, merely the Boolean sum of all ordinal forms consisting of H ordinated with itself where H is equinumerous with F. Thus in general we may let

D16. 'Nc'F' abbreviate '(G1(EH)(G Ord H,H . H Equinum F))',

and in particular then

D17a. '0' abbreviates '(F1)(EG)(F Ord G,G . G Equinum NFrm))',

D17b. '1' abbreviates '(F1(EG)(UnitFrm G . CardFrm G . F Ord G,G))',

and so on.

Let us now consider the arithmetical sum of two cardinal-unit collectives, Taylor's group of n things with another group of m things. For this let us let

D18. '(F ⊕ G)' abbreviate '(H1(EF')(EG')(F' Equinum F . G' Equinum G . (F' ∩ G') Incld NFrm . H Ord (F' ∪ G'),(F' ∪ G')))'.

Then an expression for the arithmetical sum of two cardinal numbers

D19. '(F + G)' may be defined as '(ιH . (EF')(EG')(F Eq Nc' F' . G Eq Nc'G' . H Eq Nc'(F' ⊕ G')))'.[8]

What we may now dub as Taylor's theorem (sic![9]) is forthcoming as the principle that

⊢(F)(G)(F')(G')((Coll(CardUnit) F . Coll(CardUnit) G . F' Eq Nc'F . G' Eq Nc'G) ⊃ (F' + G') Eq Nc'(F ⊕ G)).

The before-after relation among natural numbers, mentioned earlier, is merely now that between any number n and its successor (n + 1). Thus

D20. 'F Suc_{NN} G' is short for '(EF')(EG')(F Eq Nc'F' . G Eq Nc'G' . F Eq (G + 1))'.

Thus far in the technical development we have spoken only of cardinal numbers as constrasted with the *natural* numbers. The former include all numbers *of* given cardinal-unit collectives, even including transfinite ones — if such are appropriately provided for. The natural numbers, on the other hand, are the finite integers 0, 1, 2, and so, each a finite number although the number of them all is infinitely large. There would seem to be little hint in Plato of this difference, the finite cardinal integers being of primary concern. To introduce the very terms 'finite' and 'infinite' as applied to numbers requires a good deal more formal development of the theory than can be given here.

Taylor's second contention, it will be recalled, concerns the development of the real-number system in terms of Dedekindian cuts. This too may readily be carried out in the present basis, but with a good deal of additional formal material.

One can also develop here a metalinguistic theory of relations in intension, essentially as in OM or HL, by means of which we can perhaps get still closer to the spirit of the Platonic *corpus*.

Thus far we have concentrated exclusively on the theory of form and number but have neglected the status of sensible things, "concrete" individuals, and so on. The fundamental notion needed now is that of participation in the sense that a thing x participates in a cardinal form F. How can we graft on to the preceding theory a theory of individuals and of participation? A full sketch of a theory of this will not be undertaken here, where we must be content merely with a few suggestions.

Let us introduce another sort of variable 'x', 'y', and so on, for individuals and let

'x Prtcpt F'

express primitively that x participates in F. Immediately we should arrange that certain principles obtain, as follows.

In the foregoing, a good deal has been made of cardinal-unit forms and these are in effect forms determinative of individuals. This we may now make explicit by letting

'F Dtrmn x' abbreviate '(CardFrm F. UnitFrm F . x Prtcpt F)', to express that the cardinal-unit form F *determines* the individual x. It then should hold that every individual is determined by at least one form,

$$\vdash (x)(\mathrm{E}F)F \text{ Dtrmn } x,$$

and that every cardinal-unit form determines one and only one individual. To express this latter we first let

'$x = y$' abbreviate '$(F)(x \text{ Prtcpt } F \supset y \text{ Prtcpt } F)$'

in more or less familiar fashion. Then it holds that

$$\vdash (F)((\text{CardFrm } F \,.\, \text{UnitFrm } F) \supset (\mathrm{E}x)(F \text{ Dtrmn } x \,.\, (y)(F \text{ Dtrmn } y \supset y = x))).$$

Also it should obtain that

$$\vdash (x)(F)(x \text{ Prtcpt } F \supset \text{CardFrm } F),$$

so that ' . CardFrm F' can be deleted in the definiens of the definition of 'Dtrmn'. Also a kind of extensionality law should obtain that

$$\vdash (x)(y)(F)((x \text{ Prtcpt } F \,.\, y \text{ Prtcpt } F \,.\, \text{UnitFrm } F) \supset x = y).$$

In sum, according to Taylor, in another context[10], "the main ideas [in the discussion of number by Plato and in the Academy] are sound and fertile, but the formulation of them involves illogicalities. The remedy for such a situation is neither, with Aristotle, to dismiss the theory on the ground of these illogicalities, nor, like rash admires of Plato, to pretend that the illogicalities are not there. It is to reconstruct the theory with a more exact and subtle logic, and to show that its apparent paradoxes are incidental excrescences; the only way to get bad logic out of a philosophical theory is to have more logic, not less" — and good logic rather than bad, it should be added. The attempt has been made, in the foregoing, to introduce more logic even than Taylor does, and, it is hoped, a more subtle one and one closer to the spirit of Platonism, than the set-theoretic ilk used in most recent discussions of the subject.

Anders Wedberg, in his *Plato's Philosophy of Mathematics*[11], dismisses Taylor on the grounds that he "seems unduly inclined to interpret Plato's statements as anticipations of late developments within mathematics." Plato as a great anticipator has already been commented on above, and Taylor does of course take later developments fundamentally into account in his discussion. But so too does Wedberg himself. Let us recall the salient points in Wedberg's "rationalization," as he calls it, of Plato's views on number and then note how they accord or fail to accord with the reconstruction above.

Wedberg, it will be recalled, put forward several theses "on the nature of arithmetic that, ... [he believes] can be attributed to Plato." The most important of these, for present purposes, are that the "Mathematical Numbers" are characterized as follows.

"(1) They are made up of certain ideal "units" or "*1*'s". The Mathematical Number *n* is the set [!] of *N* such units: 2 is a set of two, 3 a set of three, and so on.

(2) Of such ideal units, or *1*'s, there exists an infinite supply.

(3) There is no difference between the ideal units: two such units are completely indistinguishable.

(4) An ideal unit does not contain any plurality of parts, or constituents, or characteristics: from whatever point of view we consider such a unit, it is One and One only.

(5) Of each Mathematical Number there are infinitely many copies. From the infinite supply of ideal units we may choose *N* units in infinitely many ways, and every choice gives us a representation of the Mathematical Number *N*.

(6) The elementary arithmetical notions are simple set-theoretical notions.

(7) Mathematical Numbers are the numbers studied by arithmetic. It is for them, and only for them, that the concepts of arithmetic are defined."

Here, as is the case with those of A.E. Taylor, the problems connected with justifying these contentions on the basis of the actual text are formidable, as Wedberg of course is fully aware. Even so, they may be accepted, for the moment anyhow, as containing some little glimmer of correct first approximation.

The main point here is that Wedberg makes fundamental use of the modern notion of set, just as Taylor does. Unlike Taylor, however, he reads this very notion into Plato himself. In (1) a number is a *set* of units, and in (6) the elementary arithmetical notions are "simple set-theoretic" ones. This latter especially seems highly questionable, there being no explicit set-theoretic talk in Plato at all. How simple are the set-theoretic notions for addition and multiplication, for example, in modern formulations? A casual glance at part III, Section B, of Volume II of *Principa Mathematica* — especially the definitions *110.02 of the arithmetical sum of two cardinals, *113.02 of the arithmetic product of two cardinals, and of

*116.02 of the exponentiation of two cardinals — shows them to be rather elaborate. Thus they are by no means "simple" in type-theory. But they are no less simple in set-theory, and only slightly less simple if confined to just finite cardinals. Also, as in *D16* above, the cardinal number of a collective may be construed as the sum of all ordinal pairs of forms of a certain kind, so that a closely allied notion of number is in fact available without the notion of set.

The main content of (1) is that numbers are "made up of certain ideal "units" or "*1*'s." To be "made up" of units is handled above in terms of Boolean sums. A Boolean sum is made up of the units that are included in it. The use of units here is absolutely fundamental, and Wedberg is surely correct in calling attention to it. It is natural to think of a unit as a "1", and then to confuse the two. One of the biggest problems the Greek philosophers of mathematics had to face was how to distingish the two. The number 1 is not itself a unit but is closely associated with such — more specifically, with cardinal units. And similarly, the number 2 is closely associated with two cardinal units taken together to form a sum. The exact character of this close association we have seen to be as follows. Given a sum of two cardinal units, say, the cardinal number 2 is the sum of all dyadic-ordinal forms *F* whose ordinands are some form *G* with *G*, where *G* is equinumerous with a Boolean sum of distinct cardinal-unit forms *H* and H′. Thus we may let

D17c. '2' abbreviate '(*F*1(E*G*)(*F* Ord *G*,*G* . (E*H*)(E*H*′)(CardFrm *H* . UnitFrm *H* . CardFrm *H*′ . UnitFrm *H*′ . *G* Equinum (*H* ∪ *H*′))))'.

That the units involved are "ideal" is merely to say that they are forms. Even a dual form is a unit in this sense, and similarly for triples, and so on. Thus, 2, according to the definition just given, is the sum of all ordinal forms *F* with a "dual" form *G* as its ordinands, where *G* is equinumerous with a sum of two cardinal-unit forms.

Note that Wedberg's item (2) is essentially an Axiom of Infinity, supplied here by appropriate assumptions. Further — point (3) — the ideal units are alike as regards their character as forms. The theory above distinguishes many kinds of unit forms — "individuals" or cardinal units; dyadic, triadic, etc., ordinal relational individuals of cardinal units; and so on and on. For present purposes, we consider only cardinal units. Any two cardinal-unit forms are "completely indistinguishable" as such, so to speak. In arithmetic, however, differences abound. There is only 1 table before me with 1 top, at least 4 legs, billions of atoms whirling around within it, and so on. Here the notions of being a table, a leg, a table-top, and the like, are all taken as forms comprising different kinds of cardinal units. Cardinal units are all alike *qua* being cardinal-unit forms, however, and the theory is an ontological one (as in HL), in essentially Leśniewski's sense, rather than a mereological one (as in EM). Thus, an ideal unit — Wedberg's point (4) — "does not contain any plurality of parts, or constituents."

Point (5) is that, given any Mathematical Number, there are infinitely many copies. This may be enunciated as a theorem. For a form to be "infinite" here, we may mean a form equinumerous with a form properly included in it — the so-called "reflexive" rendering of the infinite (*PM*, *124). The notion of equinumerosity needed here is an extension of that of *D15* above. Let us call it 'Equinum_2', which may be presumed defined as between expressions for relational collectives of cardinal units. A cardinal number of *F* is such an entity, it will be recalled, being a sum of all ordinal forms of some *H* with itself where *H* is a collection of cardinal units and *H* is equinumerous with *F*. Thus, our theorem is to the effect that

$$\vdash (F)(\sim \text{NFrm}\ F \supset (\text{E}G)(G\ \text{Incld}\ \text{Nc'}F\ .\ \sim \text{Nc'}F\ \text{Incld}\ G\ .\ \text{Nc'}F\ \text{Equinum}_2\ G)).$$

To be a "copy" of a number Nc'*F* here is merely to be one of the ordinal-unit forms of which Nc'*F* is a Boolean sum.

A final comment. Let us dub the foregoing system for handling forms the system F. It amounts, as already in effect noted, to a *formological* or *formtheoretic* interpretation of OM or HL. The three interpretations may be summarized as follows. In the mereological one, the values for the variables are individuals including parts of them, the units being the *minimal* parts admitted. If parts of these minimalia are to be introduced, these must be handled in terms of additional primitives. The mereological interpretation can thus be brought into accord with *physicalism*, the individuals and their parts being taken as in some physical theory. The ontological interpretation takes individuals in the sense of a Kotarbinski-like *reism* as the unit entities. If parts of them are to be recognized, this must be done in terms of additional primitives. Finally, the formological interpretation takes forms as the fundamental entities, cardinal-unit forms playing therein essentially the role of determining, but by no means being identical with, individuals. Here too, if individuals are to be accommodated as such, this is to be done in terms of some new primitive such as 'Prtcpt' for participation.

CHAPTER EIGHT

On relational couples and ordinal individuals: Peirce and Russell

> "I do not think it is possible to write the history of a science until that science itself shall have been understood, thanks to a clear, explicit, and decent logical structure."
>
> C. Truesdell

In his *The Principles of Mathematics*[1], Russell commented that "the logic of relations has a more immediate bearing on mathematics than that of classes or propositions, and any theoretically correct and adequate expression of mathematical truths is only possible by its means. Peirce and Schröder have realized the great importance of the subject, but unfortunately their methods, being based not on Peano, but on the older Symbolic Logic derived (with modifications) from Boole, are so cumbrous and difficult that most of the applications which ought to be made are practically not feasible. In addition to the defects of the old Symbolic Logic, their method suffers ... from the fact that they regard a relation essentially as a class of couples, thus requiring elaborate formulae of summation for dealing with single relations." These comments seemed cogent no doubt at the time Russell wrote, but, even so, some of them are highly dubious. The methods of the algebra of relations have not proved to be "so cumbrous and difficult" as not to be practically applicable to mathematics and elsewhere.[2] Also, it is doubtful that Peirce regarded — in the papers to which Russell refers[3] — (dyadic) relations as "classes of couples" and that the "formulae of summation" he used are any more elaborate than Russell's own methods. Nor did Russell know of mereology or the calculus of individuals, in which the only "formulae of summation" used are ordinary Boolean ones. If Peirce (and Schröder) were to have regarded unit relations or ordinal couples mereologically, as entities on a par with their relata, they could have regarded relations as Boolean sums of such couples satisfying such and such conditions. Were they to have done so, the foundations of mathematics could have stepped ahead in an important respect by almost a hundred years.

Let us explore a little these various contentions, starting more or less from scratch.

On p. 99, Russell commented — most interestingly for present purposes — that "there is a temptation to regard a relation as definable in extension as a class of couples.... But it is necessary to give sense to the couple, to distinguish the referent from the relatum: thus a couple becomes essentially distinct from a class of two terms, and must itself be introduced as a primitive idea. It would seem, viewing the matter philosophically, that sense can only be derived from some relational proposition, and that the assertion that *a* is referent and *b* relatum already involves a purely relational proposition in which *a* and *b* are terms, though the relation asserted is only the general one of referent to relatum."

Let us indeed view the matter philosophically, and in as simple and adequate a way as we can muster up. Russell thinks that this is to be done only in terms of a "relational proposition." Instead of following this well-trodden path, let us reflect upon relations themselves, rather than upon the propositions embodying them — rather mysterious entities at best be it said.[4] And let us take the notions of a *couple* seriously — in such a way as "to give sense [and order] to the couple" and thus "to distinguish the referent from relatum," to see it as "essentially distinct from a class of two terms," and, indeed, in such a way as to "be introduced as a primitive idea."

Let us first reflect upon Boolean atoms or units in mereological terms, as in OM or HL above. Let 'P' be the primitive for the relation of part-to-whole in terms of which the Boolean notions are introduced. A unit, then, relative to P, is any entity that is non-null and bears P to all non-null entities that bear P to it.[5] A unit is thus an entity not further divisible, so to speak, in terms of P. In the merological interpretation of OM, the units are the minimal entities recognized, but still need not be "individuals" in any usual sense. No, they bear P to individuals, and individuals are regarded as Boolean sums or "collectives" of these minimalia. In the "ontological" interpretation, that of HL, however, the units are taken as individuals, and collectives are then Boolean sums of individuals. In either interpretation, it is interesting to observe, we gain the effect of speaking of the membership of a unit in a class by saying that the given unit bears P to a given collective.

A collective in general, it will be recalled, is any sum of units, and in particular any sum of units satisfying such and such a condition. In the symbolism of HL, we have an Abstraction Principle, it will be recalled, that

$\vdash (x)(\text{Unit}\ x \supset (x\ \text{P}\ (y1\text{--}y\text{--}) \equiv \text{--}x\text{--}))$,

that any unit is such that it is part of a collective of units satisfying a given condition if and only if it satisfies that condition itself. This principle corresponds, of course, to a familiar kind of principle in class-theory, or even in virtual-class theory. Thus where

'$\{y \ni (\text{Unit}\ y\ .\ \text{--}y\text{--})\}\ x$' is defined merely as '$(\text{Unit}\ x\ .\ \text{--}x\text{--})$', where (as needed),

we have that

$\vdash (x)((\text{Unit}\ x\ .\ x\ \text{P}\ (y1\text{--}y\text{--})) \equiv \{y \ni (\text{Unit}\ y\ .\ \text{--}y\text{--})\}x)$.

Let 'Coll(Card) *y*' express that *y* is a collective (of cardinal individuals). Then quantifiers over classes are gained in part by

'$(y)(\text{Coll(Card)}\ y \supset \text{--}y\text{--})$'.

Let us now turn to "couples," regarded now as the units, dyadic ordinal individuals of HL. Let '*x* Ord *y*,*z*' express that *x* is the "couple" of *y* and *z* with *y* as referent and *z* as relatum. We can then let

'(*y* ord *z*)' abbreviate '$(\iota x\ .\ x\ \text{Ord}\ y,z)$',

so that the couple (*y* ord *z*) is the one and only one entity bearing Ord to *y* and *z*. Note that everything Russell requires is provided for: the "sense" of the couple in terms of the distinction between referent and relatum, the differentiation of the couple from "a class of two terms" — and also from a Boolean sum of two terms — and the introduction of it in terms of 'Ord' as a primitive notion. Triples, quadruples, and so on, can be handled by iteration, so that

'*x* Ord *y*,*z*,*w*' is short for '$(\text{E}u)(u\ \text{Ord}\ z,w\ .\ x\ \text{Ord}\ y,u)$',

and so on.

How is a couple related to its referent and relatum? Surely not in terms of part-to-whole. The couple is not a part of its relata any more than these latter are parts of it. No, the referent is the first ordinand of the couple, the relatum the second ordinand of the couple, and the couple is the *couple* of its referent and relatum — all of which is obvious enough and as it should be. (The only exception here is the null individual, which, being a part of everything, is allowed to be a part of the couple as well as of either its referent or relatum.)

It is natural now to assume couples as entities along with the units, individuals, sums of them, and so on, already availabe. They are needed in order to move on to handle "relational propositions." The very essence of relationality, we might say, resides in couples, triples, and so on, and in how we handle them. In other words, we ought to be able to develop the full logic of relations in terms of them in some suitable way. And indeed we can.

If, now, a couple (x ord y) does not have its referent x and relatum y as parts, what kinds of parts does it have? we may well ask. The answer is: it has only *itself* as a part — and the null entity. Further, it is itself non-null, not being a part of either its referent or relatum. A couple is thus both non-null and is a part of all its non-null parts. *Ergo*, couples are units. And similarly for triples, and so on. Unit entities that are not couples, triples, and so on — together with Boolean sums of such units — are *cardinal* entities, already spoken of parenthetically above.

Given couples, we can now form Boolean sums of them, and even Boolean sums of them satisfying given conditions, these latter being sums of unit couples the referents and relata of which satisfy given conditions. With all this in mind, we see how relations between cardinal entities may now be handled. A dyadic relation between cardinal entities becomes the sum of all couples the referents and relata of which are cardinal entities of which such and such a condition holds, this condition being the defining condition of the relation, so to speak.

Let us consider an example, say, the relation of being *larger-than*, as between cardinal entities, in the sense of containing more units as parts. Thus x is larger than y, in this sense, just where y bears the equinumerosity-relation Sm to some u and there is a cardinal-unit part of x not of u. Let 'CardEnt w' express that w is a cardinal entity. Our relation is then merely the sum

(1) (z1(Ex)(Ey)(CardEnt x . CardEnt y . z Ord x,y . (Ew)(Eu)(u Sm y . CardEnt w . Unit w . w P x . ~ w P u))).

And similarly for any relation between cardinals that can be expressed in the notation of the system. (On 'Sm', recall *D20* or *D20′* of OM, or *D15* of the formological system F.)

For applications of OM or HL to some non-logical subject-matter, suitable non-logical primitives may be added, together with suitable meaning-postulates characterizing them. The means are at hand to give such constants, relational or otherwise, appropriate designata within the general kind of ontology of the system.

Thus far we have considered only relations between or among cardinal entities. By parity of construction, however, we can go on to built up other kinds of relations as well. Let r_1 and r_2 be relations between or among cardinals — of which (1) above is an example. Consider now the couple (r_1 ord r_2), with r_1 as referent and r_2 as relatum. This couple is also a unit — as all couples are — and thus we can form the sum of all units of this kind the referents and relata of which satisfy a given condition. Such a sum is a dyadic relation between relations between or among cardinal entities. And so on, for all manner of relations between or among relations of all manner of complexity.

In all of this, we have gone way beyond anything envisaged by Russell. We have merely been following a suggestion that he put forward only to push aside and move ahead in a quite different way, as we have noted. Why did Russell not follow through on his suggestion concerning couples? The very next sentence on p. 99, directly following the quote above, gives us the answer. "There are, in fact," he wrote, "concepts such as greater, which occur otherwise than as terms in propositions having two terms; and no doctrine of couples can evade such propositions." Russell thought it "therefore more correct" to take another course in handling relations. Russell does not say, however, it should be observed, that no doctrine of couples *could* be formulated to handle occurrences of relational terms otherwise than as terms in propositions [sentences] having two terms." In the foregoing, we have seen how this may be done. Relations can occcur as referents and relata of suitable relations of higher order, so to speak, such relations also being handled in terms of a general theory of couples, triples, and so on. There is thus no neglect here of sentences in which expressions for relations can occur in all the ways required in a general theory.

In all of this discussion, there has been no talk of classes, of classes of classes, and so on — other than virtual ones. Lovers of class-theory may object that a couple, after all, *is* merely a class of classes in disguise. Thus, where {α, β} is the class whose only members are α and β — no matter what kind of entities α and β are — one might argue that (x ord y) *is* merely the class of classes

$$\{\{x\}, \{x,y\}\},$$

in the manner essentially of Wiener or Kuratowski.[6] Hence, one might say, the theories of OM and HL are merely a kind of class-theory in disguise. Such an attitude would seem to be rather obstinate and one-sided, it should be noted, and would seem to presuppose that the class-theoretic way of viewing couples is the one and only one. It would be like saying that identity is always to be thought of class-theoretically, that is, that '$x = y$' means that y is a member of every class that x is, and that therefore if we take '=' as a primitive in some system without classes, that system contains a class-theory in disguise. And similarly if '=' is defined in terms of the primitive predicates of the system in the manner of Hilbert-and-Bernays.[7] Any such contention would seem rather ill-judged, and the person who made them we would tend to think rather class-theoretically arrogant, or at least brain-washed, and rather intolerant of alternatives. And he would be committing the worst of methodological sins, according to Peirce, that of blocking the road of inquiry — more particularly, of inquiry into alternatives.

The point of the theory of couples, as incorporated in OM or HL, is that with their help we can achieve what otherwise can be done class-theoretically. This is methodologically interesting as a step toward economy, and helps to free us from the straight-jacket of the class-theoretical point of view, which had become in recent years the one and only "received system, considered as a standard of reference from which appeal is denied." One need not find anything objectionable in set- or class-theory in order to lament the organization-man kind of obsequiousness to it that prevents interest in and tolerance of alternatives.

It has been said of OM that an explanation is needed "of what makes entities abstract, and what are the features of sets that ... [are found] objectionable." And also that an account "should be provided to justify the claim that ... ordinal individuals are less objectionable than sets." Note how misguided and inaccurate both of these demands are. No claim was made that ordinal individuals are less "objectionable" than sets or classes. Nor were ordinal individuals claimed to be either more or less "abstract" or "concrete" than sets, classes, individuals, or what have you.[8]

Russell's comments, quoted at the beginning, were first published in 1903, long before *Principia Mathematica* and the Zermelo axiomatic set-theory came on the market. Boolean theory was already available, however. Boole himself seems to have shown little explicit interest in relations as such, although probably more than is commonly supposed. Russell commented, again in the important passages on pp. 23 ff., that "the calculus of relations is a more modern subject than the calculus of classes. Although a few hints for it are to be found in De Morgan, the subject was first developed by C.S. Peirce." This comment is surely not just to De Morgan, as has in effect been urged elsewhere.[9] We should also query Russell's remarks, already quoted, concerning Peirce's "cumbrous and difficult" methods, concerning whether in fact he regarded "a relation as essentially a class of couples," and concerning the "elaborate formulae of summation for dealing with single relations."

In the first of the two papers of Peirce to which Russell makes reference, the notation '(A:B)' is introduced to "denote individual relatives." An "individual" for Peirce is essentially the same as a unit cardinal of HL. Peirce comments that "every term [or rather collective] may be conceived as a limitless logical sum of individuals, or as a limitless logical product of simples." (A "simple" for Peirce is the negative of an individual. In the notation of HL, we may let

> 'Simple x' abbreviate '$(\sim \mathrm{U\,P}\,x\,.\,(y)((\sim \mathrm{U\,P}\,y\,.\,x\,\mathrm{P}\,y) \supset y\,\mathrm{P}\,x))$'.[10]

This is a kind of "dual" to the definition of 'Unit.)

In Peirce's symbolism, where a is a "term" and the A_i's are individuals, we have that

(P1) "$a = \mathrm{A}_1 + \mathrm{A}_2 + \mathrm{A}_3 + \mathrm{A}_4 + \mathrm{A}_5 + \text{etc.}$

(P2) $\bar{a} = \bar{\mathrm{A}}_1 \times \bar{\mathrm{A}}_2 \times \bar{\mathrm{A}}_3 \times \bar{\mathrm{A}}_4 \times \bar{\mathrm{A}}_5 \times \text{etc.}$

It will be seen that a simple is the negative of an individual." Yes, but from the definitions of 'simple' and 'individual', not from these two formulae.

Given any individuals A and B, we may form dual relatives (A:A), (A:B), (B:A), and (B:B). And similarly, triples, given any three individuals. Given any n distinct individuals there are 2^n triples of them. And similarly for simples. And similarly there are dual relatives with an individual as one component and a simple as the other. Thus (A:$\bar{\mathrm{B}}$), ($\bar{\mathrm{A}}$:$\bar{\mathrm{B}}$), and so on, are also recognized as dual relatives.

Can relatives involving simples be construed wholly in terms of relatives involving only individuals? Yes, of course, although Peirce does not state this crucial fact. Let (A:A), (A:B), (B:A), (A:C), (B:B), (C:A), (A:D), (B:C), (C:B), (D:A), and so on, constitute an enumeration of all the individual dual relatives — note the *Diagonalsverfahren* involved. Then any $\bar{A}$, where A is an individual, is itself merely a sum of individuals, say $A_1 + A_2 + \ldots$. How then are we to construe, say, $(A:\bar{A})$ or $(A:(A_1 + A_2 + \ldots)$? Peirce does not tell us, but he seems to need a kind of distributive law that

(P3) $(A:\bar{A}) = (A:A_1) + (A:A_2) + \text{etc.}$,

and, more generally, that

(P4) $\bar{B} = (B_1 + B_2 + \text{etc.}) \supset ((A:\bar{B}) = (A:B_1) + (A:B_2) + \text{etc.} \,.\, (\bar{B}:A) = (B_1:A) + (B_2:A) + \text{etc.})$

And similarly for triple relatives, and so on.

With these formulae in mind we can easily see that, as Peirce tells us, where ∞ is "the logical sum of all the relatives" in the enumeration,

(D5) $\infty = (A:B) + (\bar{A}:B) + (A:\bar{B}) + (\bar{A}:\bar{B})$.

He also states that

(P6) $\infty = A + \bar{A}$,

so that the logical sum of all dual relatives must be the same as the logical sum of all individuals. And similarly these logical sums must be the same as the logical sum of all triple relatives of individuals, for we are also told that

(P7) $\infty = (A:B:C) + (\bar{A}:B:C) + (A:\bar{B}:C) + (A:B:\bar{C}) + (\bar{A}:\bar{B}:\bar{C}) + (A:\bar{B}:\bar{C}) + (\bar{A}:B:\bar{C}) + (\bar{A}:\bar{B}:\bar{C})$.

Prima facie, looks as though Peirce is equating here three quite different kinds of sums. And indeed he is, for we are told that "it is needless to distinguish the dual universe, the triple universe, etc., because, by adding a perfectly indefinite additional member to the system, a dual relative may be converted into a triple relative. Thus for *lover of a woman*, we may write *lover of a woman coexisting with anything*. In the same way, a term of single reference is equivalent to a relative with an indefinite correlate; thus, *woman* is equivalent to *woman coexisting with anything*. Thus we shall have

(P8) $A = (A:A) + (A:B) + (A:C) + (A:D) + (A:E) + \text{etc.}$

(P9) $(A:B) = (A:B:A) + (A:B:B) + (A:B:C) + (A:B:D) + \text{etc.}$"

Strictly we should distinguish here an individual name from a general noun such as 'woman', and we shall wish to say very different things about them. Peirce does not introduce duals such as (woman:man) or even (woman:Helen) or the like. No, the theory centers around individuals and simples, duals, triples, and so on, comprised of them, and logical sums.

Thus the example of the "equivalence" (in some sense) of 'lover of a woman' with 'lover of a woman coexisting with anything' seems ill-chosen. Let 'w' represent here 'woman' and 'c' 'coexisting with anything'. Then 'woman coexisting with anything' would be best represented by 'w × c' and not by '(w:c)', which is not — immediately anyhow — a dual relative of *individuals*. Peirce seems to be confusing here two very different kinds of things, a Boolean *product* of w and c with dual relatives of individuals.

Another problem arises. Peirce should have noted that, if (P8) and (P9) are assumed, then for essentially the same reason it should be assumed also that

(P10) A = (A:A) + (B:A) + (C:A) + etc.,
(P11) A = (A:A:A) + (B:A:A) + (A:A:B) + (B:A:B) + (C:A:A) + (A:A:C) + (C:A:C) + etc.

and so on, for all quadruples, quintuples, and so on.

To equate an individual with sums of dual relatives, sums of triple relatives, and so on, seems somewhat counterintuitive. (The theory of relatives, it would seem, should be *well-grounded* in the individuals concerned, and wholly determined by them.) And, anyhow, what advantage is to be gained by blurring these all together in this fashion? And does it lead us into technical difficulties as the theory develops? Yes, it seems, it does.

Peirce is not one to have worried much about the full systematicity of what he does — of the whole vocabulary needed, of the primitives, of the notions of 'formula' and 'term' needed, of precise definitions of all non-primitives, of the axioms, and of the rule needed to go from axioms to theorems. The very notion of a "proof" or "derivation" in Peirce's theories is thus rather obscure. Even so, we may move ahead as follows, as he himself frequently does.

Let us bring in the '—<' which plays the role in Peirce's theory that 'P' does in HL or EM. And identity for Peirce may be handled in terms of '—<', so that *a* = *b* if and only if *a* —< *b* and *b* —< *a* — recall 3.173. Thus, from (P8) we can infer that (A:A) —< A, (A:B) —< A, and so on. And similarly (A:A:A) —< A, and so on. And of course each of the summands of (P10) and (P11) also bears —< to A.

But A is an *individual*, it should be recalled, so that A in turn bears —< to everything non-null that bears —< to it. Each of the summands of A is non-null, surely, although we are not explicitly told so. Hence A bears —< to each and every one of them. Let us consider now any distinct individuals A and B and observe that from what has just been said, we have that

$$A \multimap (A:B) \text{ and } (A:B) \multimap B.$$

But —< is surely transitive, so that we have also that

$$A \multimap B.$$

Thus for any two individuals, one bears —< to the other. By a similar reasoning, it can be shown that the other bears —< to the one. Thus Peirce's theory, as developed in this *American Journal of Mathematics* paper of 1880, is such as to countenance just one individual — contrary surely to its intent.

Another troublesome feature of Peirce's discussion is the vague use of 'etc.' in the formulae. To make this use acceptable, to mathematize it in a precise notation, some device such as that of the unit summation of HL seems needed. Peirce's blurring of individuals, dual relatives of them, and so on, is then captured by the formula of HL that

$$x = (y1(\mathrm{E}z)(\mathrm{E}w)(y \text{ Ord } x,z \text{ v } y \text{ Ord } z,x \text{ v } y \text{ Ord } z,x,w)).$$

But this formula is surely not acceptable and would play havoc with mereology, just as (P10) and (P11) do with the Peircean form of theory.

In place of unit summation or something similar, Peirce uses the unanalyzed '+ etc.' notion, as we have observed. This use involves two other undesirable features. The first is that all the individuals A, B, and so on, are presumed enumerable and, moreover, nameable. How many individuals are there? Perhaps only countably many, perhaps more. In any case, the logic of relations should not have as a presupposition that there are only countably many. Nor should it presuppose that each is nameable or, more specifically, actually named, as with Peirce. Note also that if there are non-countably many individuals, then several of Peirce's formulae must themselves be of infinite length. Languages containing expressions of infinite length have been studied a little, but the general opinion seems to prevail that they are of only minor mathematical interest and of no help in the study of natural language. In any case, the logic of relations should not have to depend for its formulation upon the use of such languages.

We have already observed that Peirce does not distinguish the "universe" of individuals from that of dual relatives, nor neither of these from that of triple relatives, and so on. We have also seen that not to do so leads to unfortunate consequences. It is of interest to observe here how these consequences are avoided in OM (and, *mutatis mutandis*, in HL). In OM, where W is the total universal individual, and V_2, V_3, and so on, are the universal dyadic relation between cardinal individuals, the universal triadic relation among cardinal individuals, and so on, we have that

$$\vdash (\sim W P V_2 . V_2 P W . \sim W P V_3 . V_3 P W . \ldots . \sim V_2 P V_3 . \sim V_3 P V_2 . \sim V_2 P V_4 . \sim V_4 P V_2 . \ldots . \sim V_3 P V_4 . \sim V_4 P V_3 . \ldots),$$

not all units being dyadic relations between cardinal individuals but all such relations being part of W, no dyadic relation between cardinal individuals being a triadic one, and so on.

It should be observed that Peirce is no longer concerned with individual relatives in the second *Americal Journal of Mathematics* paper, of 1885, to which Russell made reference. Here the whole *venue* has changed radically. In this paper Perice was in full command of the logic of quantification in essentially the manner of the Whitehead-Russell treatment in *Principia Mathematica*. In the *Principles of Mathematics*, as we have noted, Russell lumps these two papers together indiscriminately, apparently not aware of the difference. In this latter paper, surely, individual relatives are not regarded as "classes of couples" nor are "elaborate formulae of summation for dealing with single relations" required. Nor are relatives regarded as "classes of couples" in the earlier paper, as we have noted, but rather as Boolean sums of them. However, in that paper the formulae of summation are rather elaborate, and, moreover, some of them are incorrect. Also it is not clear that the theory is developed extensively enough to handle relations between and among relations in full generality, including relations of higher order. Some of Russell's strictures seem therefore, in the main, justified. But his positive suggestions concerning couples have had to wait eighty years for a proper airing.

In his valuable review of Vol. III of Peirce's *Collected Papers*[11], Quine pointed out many items of the utmost importance for a proper understanding of Peirce's work in logic. It is unfortunate that this review has not been paid attention to by Peirce scholars. A careful appraisal of it is very much overdue. A minor point that Quine mentions is the need for the inner parentheses in (P8) and allied formulae that are missing in the printed text. It is interesting, however, that Quine does not call attention to the more fundamental error pointed out above, lurking in (P8) together with the other formulae and the definition of 'individual'. This error is fundamental and seems to infect all of Peirce's work on the logic of relatives until he had come to terms with the quantifiers in the paper of 1883[12] and the *American Journal of Mathematics* paper of 1885.

CHAPTER NINE

On ordinal nominalism: Goodman, Eberle, and Hempel

"Nemo potest nudo vestimenta detrahere."

Let us reflect for a moment of Goodman's much discussed *principle of nominalism*, to the effect that there is "no distinction of entities without a distinction of content."[1] More specifically, "for a nominalistic system no two different things have the same atoms: only from different atoms can different things be generated; all non-identities between things are reducible to non-identities between their atoms." The use of 'content' seems vague and Goodman does not define it. Of course 'atom' is vague also unless suitably clarified. Goodman does specify it somewhat by bringing in a "generating" relation *R* between "things" in such a way that a thing *x* is an atom relative to *P* if and only if nothing bears *R* to *x*.

The units of OM and HL, as discussed in V above, play the respective roles therein of Goodman's atoms. They are the ultimate entities out of which all others may be generated. And by 'content' or 'content of *x*', let us mean the unit parts of *x*. What then is the "generating" relation — or, better, the generating virtual relation? Merely

(1) $\{xy \ni (\sim x \mathrm{\,P\,N} \,.\, x \mathrm{\,P\,} y \,.\, \sim y \mathrm{\,P\,} x)\}$

or, regarded as a relation of OM (or HL) itself,

$$(z1(\mathrm{E}x)(\mathrm{E}y)(\sim x \mathrm{\,P\,N} \,.\, x \mathrm{\,P\,} y \,.\, \sim y \mathrm{\,P\,} x \,.\, z \mathrm{\,Ord\,} x,y)).$$

Let Gen be the virtual relation (1). Then clearly

$$\vdash (x)(\mathrm{Unit}\ x \equiv \sim (\mathrm{E}y)(y \mathrm{\,Gen\,} x),$$

so that "*x* is an atom relative to *R* [i.e., Gen] if and only if nothing bears *R* to *x*." Likewise

$\vdash (x)(y)((\sim x = y \, . \sim x \, \mathrm{P} \, \mathrm{N} \, . \sim y \, \mathrm{P} \, \mathrm{N}) \supset \sim (z)(w)((\mathrm{Unit}\ z \, . \, \mathrm{Unit}\ w \, . \, z \, \mathrm{Gen}\ x \, . \, w \, \mathrm{Gen}\ y) \supset z = w)),$

so that "only from different atoms can different things be generated;" and

$\vdash (x)(y)((\sim x = y \, . \sim x \, \mathrm{P} \, \mathrm{N} \, . \sim y \, \mathrm{P} \, \mathrm{N}) \supset (\mathrm{E}z)(\mathrm{E}w)(\mathrm{Unit}\ z \, . \, \mathrm{Unit}\ w \, . \sim z = w \, . \, z \, \mathrm{Gen}\ x \, . \, w \, \mathrm{Gen}\ y)),$

so that "all non-identities between things are reducible to non-identities between their atoms." *Ergo*, the principle of nominalism is satisfied in OM and HL where the Units are regarded as the atoms relative to the generating virtual-relation Gen.

But of course much more is dreamed of here than in Goodman's philosophy, namely, the ordinal individuals. These are regarded as units or atoms, with no non-null "parts" other than themselves. Recall that where *x* Ord *y*,*z* neither *y* nor *z* are parts of *x* unless they are null. Nor is *x* a part of *y* or *z*. We may say that *x* and *z* are *ordinands* of *x*, or *segments* of *x*, or something of the sort — but this is a very different matter and subject to very different logical laws.

Hempel, in his review of Goodman's *The Structure of Appearance*[2], commented on the principle of nominalism as follows. "The insistence ... that two entities cannot be different if they have the same content does not appear to me to be convincing so long as the notion of content is used in ... a vague and unspecified manner. Thus, e.g., the idea of different sequential arrangements of a given finite number of individuals — or, as a special case, the formation of two different ordered couples from two individuals — would seem to be quite clear, and yet we have here apparently several different entities with the same content." Well, Hempelian clarity is a high desideratum, surely, and what is "quite clear" for Hempel should be clear enough for the rest of us — provided the "couples" involved are characterized by exact rules themselves having the requisite clarity, simplicity, fruitfulness, and so on.

Goodman's rejoinder to Hempel, in "A World of Individuals" (p. 24), seems quite beside the point. "Normally," he writes, "we no more conclude that we describe different entities when we name two people in different order than we conclude that a house from top to bottom and the house from bottom to top are different entities, or that the capital of Massachusetts and the largest city in New England are different things. We do not take the varied histories of the Battle of Bull Run as recounting different occurrences. In daily life a multiplicity of descriptions is no evidence for a corresponding multiplicity of things described." Certainly not, but we must not confuse a thing under a description with an ordinal individual as such. The pair *Washington*, *Lincoln* — to use Goodman's example — is not *Washington* taken under any description at all. As a matter of fact, however, Washington-taken-under-a-description-*a* can be regarded as the ordinal individual

(Washington ord *a*),

where *a* is a one-place predicate describing him. However, Hempel's point, as we have seen, does not rest upon this confusion in any way.

It should be observed also that Goodman's reply here is such as to rule out the purported treatment of inscriptional concatenation in the Goodman-Quine "constructive nominalism." If 'C *abc*' expresses that *a* is the concatenate of inscription *b* and *c*, *a* here must be distinguished from a *d* where C *dcb*. It is essential to be able to distinguish these for any viable inscriptional syntax. Here both *a* and *d* are a special kind of ordinal individual, and in fact may explicitly be handled as such.[3] Any objection Goodman might put forward against the notion of an ordinal individual could thus readily be transformed into an objection against the very handling of concatenation that is so fundamental to the kind of nominalism he espouses.

It has often been said, and justly so, that the greatest weakness of Goodman's nominalism — an interesting and important view in many respects, be it said — is that it is not able to accommodate mathematics in any deep sense. Not to do so is of course to give hostage to fortune. It cannot provide a syntax for it, as we have just noted, let alone a system in which mathematics may be expressed or a semantics in which it may be interpreted. Many other philosophies have also been shipwrecked in this way — the very kinds that Goodman most despises. If nominalism is broadened, however just to the extent of admitting ordinal individuals, in the manner of V above, a foundation for mathematics is forthcoming, as we have seen — on, of course, suitable assumptions. To admit ordinal individuals, moreover, is not to open the flood-gates to the morass of undesirable entities to which Goodman objects. This is the important point. Ordinal nominalism is still a severe view, but one that does not close its doors to the stupendous achievements of mathematics throughout its long history.

In spite of all this, it is unlikely that the general notion of an ordinal individual can be brought into accord with Goodman's brand of nominalism. Many philosophers regard this latter as too narrow anyhow[4], and in need of a considerable extension in order to "make sense" — Goodman's very phrase — of the languages of mathematics and of the sciences. *Tant pis*, then, for the narrow nominalism; let us welcome in its stead the more commodious kind of *theoretical* or *ordinal* nominalism that can perform these tasks as efficaciously as any set-theory can, and, moreover, in such a way as to satisfy Goodman's principle of nominalism suitably construed.

In his *Nominalistic Systems*[5] Rolf Eberle has discussed several ways in which "sequential individuals" might be introduced within a quasi-Goodmanian framework. He comments (p. 82) that "in as Goodman appeals to everyday instances of things, rather than to a philosophically refined notion of an individual, his inability to discover such things [as sequential or ordinal individuals] is somewhat surprizing. Whenever people speak of cloth patterns, flower arrangements, fleet formations, planetary systems, stellar constellations, clusters of berries, sentences, organisms, and so forth, they are naturally taken to speak of things which can differ just with respect to the arrangement of their parts." Yes — and *a forteriori*, surely, when we reflect upon the sophisticated modes of parlance in the theoretical sciences.

Eberle goes on to discuss four ways in which sequential individuals might be introduced without losing sight of the fundamentals of the nominalist point of view. The one that he thinks the "most advantageous" is "the only one among the four which justifies any hope that relations can be characterized in terms of sequential individuals." In this method, a triadic relation *L* is introduced primitively in such a way that '*Lxyz*' is comparable in some respects to '*y* Ord *x*,*z*' in OM or HL, but with important technical differences. In particular the "links" he admits are only as between "atoms" and the method is apparently not able as it stands to handle relations between or among individuals in general, nor among relations of higher order — and therewith the foundations of mathematics. Also Eberle makes fundamental use of a model-theoretic (and hence set-theoretic) metalanguage for the presentation of his view, such a metalanguage not satisfying even remotely the constraints of any brand of nominalism whatsoever. "We have not even begun," as he himself notes (p. 198), "to solve ... problems which beset 'metanominalism'." Even so, the theory of "links" does appear to have some family resemblance to EM and HL and is perhaps worthy of further development.

Finally, let us ask: Why is it that ordinal or sequential individuals are needed fundamentally for a viable, theoretical nominalism that takes mathematics and the sciences seriously? A suitable answer to this question is forthcoming by altering ever so slightly some valuable comments of Hempel with respect to the languages of the sciences. "In order to obtain theories of great precision, wide scope, and high empirical confirmation," he writes[6], "science has ... evolved, in its different branches, comprehensive systems of special concepts [expressible in terms of relations], referred to by technical terms. Many of these concepts [or relation] are highly abstract and bear little resemblance to the concrete concepts [names, Russellian descriptions, and predicates] we use to describe the phenomena [and individuals] of our everyday experience. Actually, however, certain connections [or relations] must obtain between [the objects of] these two classes of concepts; for science is ultimately intended to sytematize [and thereby interrelate] the data [and objects] of our experience, and this is possible only if scientific principles, even when couched in the most esoteric terms [usually for relations of higher order], have a bearing upon, and thus are conceptually connected [or related] with, statements reporting in "experiential terms" available in everyday language what has been established by immediate observation. Consequently, there will exist certain connections [or relations] between the technical terms [and their designata] of empirical science and the [items of the] experiential vocabulary [and their designata]; in fact, only by virtue of such connections [or relations] can the technical terms [and principles] have any empirical content." In fact, such terms and principles cannot even be *stated* other than by admitting expressions for these various connections or relations as part of the scientific and metascientific vocabulary. In short — no individuals, no observables; no relations, no principles; no individuals or relations, no science; no science, no philosophical intelligibility.

What does the "world" look like, according to ordinal nominalism? Again, a paraphrase of a comment by Hempel (p. 36) will be helpful by way of an answer. Hempel writes that a "scientific theory might ... be likened to a complex spatial network: Its terms are represented by the knots, while the threads connecting the latter correspond, in part, to the definitions and, in part, to the fundamental and derivative hypotheses [or principles] included in [and constituting] the theory. The whole system floats, as it were, above the plane of observation and is anchored to it by rules of interpretation. These might be viewed as strings which are not a part of the network but link certain points of the latter with specific places in the plane of observation. By virtue of these interpretive connections, the network can function as a scientific theory: From certain observational data, we may ascend, via an interpretative string, to some point in the theoretical network, thence proceed, via definitions and hypotheses, to other points, from which another interpretive string permits a descent to the plane of observation."

A "knot" here, it should be noted, represents a term for either an observable or a theoretical construct. Let us now "ontologize" Hempel's picture by letting the knots represent only cardinal individuals (in the sense of HL), only some of which are observables. Given any two cardinal individuals *y* and *z*, a thread connects them (with an arrowhead indicating the direction from *y* to *z*) because there is an *x* such that *x* Ord *y*,*z* according to the postulates. And similarly threads connecting threads, if the postulates warrant such. And so on, resulting in a vast network of knots and threads of a very complicated kind. Even so, this picture does not even begin to capture the immensity of the "world", for all manner of Boolean sums, products, and negatives are also postulated to exist, which we have no convenient way of picturing here. Also the world or cosmic entity U and the null entity must be given a place. For the ontology of ordinal nominalism, then, we need a vastly more complex array than Hempel's, an array, namely, for all the values for the variables of the system, whether named or not.

If we assume now that some of the individuals are "observables," the "strings" tying them to the plane of observation may be described essentially as Hempel does. However, it should be noted that the observables constitute only a very small fragment of the ontology here. The greater part of the cosmos floats about with no strings attached — i.e., directly, and if at all perhaps only very remotely indeed. Such is the cosmos if the high reaches of mathematics are to be taken seriously, as they most surely must in any attempt to characterize the languages of science.

One final comment. Goodman's world of individuals is comparable to that of Leibniz's monads in being "windowless." Such relations as Goodman allows may be handled merely as virtual ones and not therewith as values for variables. In the ontology of ordinal nominalism, on the other hand, the cardinal individuals are more like Whitehead's "actual entities" with their windows open to the entire cosmos, taken both distributively and collectively. Each entity is threaded with each other entity, so to speak, including its contemporaries — unlike Whitehead's in this respect — and these threads are in turn threaded with other threads and with other entities of all sorts, and so on and on. Again, nothing less than this is adequate if mathematics is to be given its proper setting in the modern theory of relations. We live, not in a "world of individuals" as with Goodman, but rather in a structured world-of-individuals-and-relations-in-indissoluble-union-with-one-another.

CHAPTER TEN

On sets and individuals: Hao Wang

Mathematica brevis, logica longa.

A certain philosophical suspicion surrounds the notion of set, and hence the notion of a domain of sets as a part or whole of an ontology. An ontology consists of the entities taken as the fundamental or basic ones, or as the values for variables in a given system, if the system contains variables. Let us explore a little, in the present paper, the grounds of this suspicion against sets. Of course some suspicion may surround any ontology at all, quite irrespective of whether sets are included in it or not. That they *must* be included within the ontology of any theory of serious philosophical or mathematical interest is the prevailing view and is commonly dubbed 'Platonic realism', but quite erroneously so as we have seen in VII above.

The truism 'No individuals, no sets' is so elementary and obvious as not to have invited comment in the philosophical literature. Sets, other than the null or empty set, are such as to have something as members. Some of these members may be entities other than sets; such entities are commonly called 'individuals' or '*Urelemente*'. Sets themselves are such, then, as significantly either to be members or to have members, whereas individuals may be members but cannot significantly have members. If there are no individuals, the only set left is the null or empty set, which could not even be described as 'empty' if there were no individuals of which it is empty. Without individuals, the empty set appears at best to be a dubious, *als ob* kind of thing. In a real sense, then, it would seem, individuals are more fundamental than sets, and, in some fashion or other and howsoever construed, are to be recognized as entities in any scientific or philosophical scheme worth the bother.

In a perceptive recent discussion[1], Hao Wang has called attention (p. 22) to the "privotal position of sets" in logic and the foundations of mathematics. "That we have an adequate formal system of pure logic (the completeness problem of the first order logic) depends for its very formulation on the concept of arbitrary sets. And we have unique (categorical) characterizations of natural numbers and real numbers by axiom systems only if we use second order theories, i.e., presuppose a fixed comprehensive concept of arbitrary sets of these numbers. Thus we see again and again that we appeal to a concept of set in order to justify absolute results in other areas. On the other hand, we do not have a similarly adequate characterization of sets.... Another fascinating feature about set theory is the apparent paradox that doubts about its foundations are widespread yet we are capable of acquiring very strong intuitions to arrive informally at correct, interesting, and coherent concepts and theorems. More, philosophical problems on mathematical objects and on the inner resources of a given branch of knowledge come to a specially conspicious focus in the examination of set theory."

What precisely is a set? we may ask. "A set is a collection of previously given objects," Wang writes (p. 181); "the set is determined when it is determined for every given object x whether or not x belongs to it. The objects which belong to the set are its members, and the set is a single object formed by collecting the objects together. The members may be objects of any sort: plants, animals, photons, numbers, functions, sets, etc. According to ... [this] iterative concept, a set is something obtainable from some basic objects (such as the empty [or null] set, or the integers, or individuals, or some other well-defined urelements) by iterated application of the rich operation 'set of' which permits the collecting together of any multitude of 'given' objects (in particular, sets) or any part thereof into a set. This process includes transfinite iterations. For example, the multitude of sets obtained by finite iterations is considered to be itself a set."

This "iterative" notion of set, Wang notes, "involves at least four difficult ideas: the idea of 'given', the idea of collecting together, the idea of 'part' or subset, and the idea of iteration." But not, note, the fundamental notion of 'object' or, equivalently, 'individual' or 'Urelement', which Wang presumes admissible independently of all talk of sets. Without some objects, or non-sets, however, Wang's charaterization of 'set' could not even be stated except where there is only one "basic" object, the null set.

Wang does not go on to defend his use of the "difficult ideas" mentioned, and his characterization does not tell us much about what sets are *as such* so much as how we arrive at them in suitable epistemic acts. The objects are "given" somehow and then we go about "collecting" them. But what then is the collection once the objects are collected? This we are not told, other than that it is "single object" formed by use of the "rich" operation 'set of'. None of these explanations have clarificatory value without our already understanding the very notion 'set of'.

Suppose we have an operation of Boolean summation of some kind upon the individuals admitted. We could then repeat one by one Wang's points for such sums. A sum is a collection of individuals; it is "determined" when it is determined for any given individual x whether or not x belongs to it in the sense of being a summand of it. The individuals that belong to the sum are just its individual summands, and the sum is a single object formed by collecting the individuals together. Further, the notion of sum is "iterative" in the sense that sums of sums may be formed by means of an operation 'sum of' — less "rich" than 'set of', be it noted — and this may be extended to infinity. Of course, the technical handling of sums will be very different from that of sets, so that the analogy between the two is not to be pressed too far. The point here is merely that Wang's informal characterization of 'set' does not go far enough to differentiate between 'set' and 'sum' without presupposing the very notion 'set' that it purports to elucidate.

Wang tells us nothing about the theory of acts needed for his characterization. Some suggestions toward such a theory have been given elsewhere[2], and are perhaps worth further developing. We could let, for example,

'p $\text{Clasfy}_e\ x,a$'

express that, in some intentional act e, person p classifies x as having the one-place predicate a apply to it. The result of all such acts, by p, would give the notion of the extension of the predicate a for p. This "result" would be something like a set, but with suitable restrictions built into it in view of the principles or meaning-postulates that would be needed to characterize 'Clsfy' as a pragmatical primitive.

Recall that Wang admits that we do not have "an adequate characterization of sets" of the kind that the use of sets provides "in order to justify absolute results in other [mathematical] areas." This is a strong indictment of set-theory of which more should be made. Doubts about "its foundations are widespread" and have become even more so in the 10 years since Wang's book was written. Our "intuitions" about it are now less strong than then, and it has always been dubious "that we are capable of arriving at correct, interesting, and coherent concepts and theorems" about it. Leśniewski, for example, could find "no coherent meaning" to the 'ε' of Zermelo and hence to no statement, mathematical, philosophical, or other, made by means of it. Similar grumblings have been made over the years by others.

Let us recall also the contention that "philosophical problems on mathematical objects and on the inner resources of a given branch of knowledge come to a specially conspicious focus in the examination of set theory." We should distinguish here the examination of set-theory from its use in carrying out an examination. Not all examinations need make use of a set-theory. It is true that if we view mathematical objects as set-theoretical ones, they take on the very focus of that theory itself — perhaps a blurred rather than a clear one. If mathematical objects are not taken set-theoretically — but *sui generis* or in some other, perhaps mereological or homological way — they also may be brought into conspicuous focus, clear or blurred. And, similarly, when we consider the "inner resources of a branch of knowledge." Set-theory may or not figure among these resources. It is important to know whether or not it does, if only to "keep tabs on our tools," to use Quine's phrase. It usually turns out that only very small portions of set-theory are needed, in most applications, and often even these may be dispensed with altogether. It is in fact remarkable how little set-theory is needed in extra-mathematical areas of inquiry, and the more one makes the effort to do without it the more one succeeds.

Wang notes that "we understand this [maximally iterative] concept of set sufficiently well to see, after some deliberation, and in some cases even a great deal of deliberation, that the ordinary axioms of set theory are true for (or with respect to) this concept, and to be able to extend these axioms by proposing different axioms and recognizing some of them to be true for (or with respect to) it." Of course — but the same may be said for many another "concept." Throughout the ages human beings here become expert at convincing, or even brainwashing, themselves into believing all manner of statements about concepts of a highly dubious character. Intellectual history is mostly the history of "the prevalence of humbug," to use Max Black's felicitous phrase. At first blush one thinks that of course the history of mathematics is an exception to this; but alas, this is not the case. And anyhow, Wang is not concerned with objective or even intersubjective truth here, but only with subjective truth relative to given persons and a given concept. This concept may be stated in terms of a primitive 'ε' for the relation of membership relative to a domain of sets howsoever taken. To "see that the axioms are true with respect to this concept" here should be construed merely in the sense that the person *p* — the mathematician or "creative subject" doing mathematics — accepts that the axioms are true of 'ε'. In saying this, we are merely reporting a subjective fact about *p*, and saying nothing whatsoever about the "truth" of the axioms independent of *p*'s "seeing." The notion 'accepts as true' here is a pragmatical, not a semantical, one.

Very well, then, let us let the mathematician have his iterative concept of set. And let him, as is now common, admit no *Urelemente* but only the empty set as the "basic" object. In doing this, mathematics is completely cut off from the world of actuality and allowed to float in thin air. The concept of 'set' employed seems highly artificial and *ad hoc* and needs a considerable epistemological founding for its intelligibility and a metaphysical one for the ontic status of its objects.

Wang points out (p. 187) that "this iterative concept of set is of course quite different from the *dichotomy* concept [italics added] which regards each set as obtained by dividing the totality of all things into two categories (viz. those which have the property [— what property? —] and those which do not). Following Gödel, one may speak of the two concepts as the mathematical versus the logical." This characterization of the dichotomy concept of 'set' is not very clear. Neither 'set' nor 'category' are in any strict sense words of logic. The theory of truth-functions and quantifiers (or its surrogate in relational algebra) does not contain these notions in any guise. At best, then, the dichotomy concept cannot be a "logical" concept, requiring for the very statement of what it is a mathematical word 'set' and a metaphysical one 'category'.

Gödel's own statement, incidentally, of the dichotomy notion contains 'existing' between 'all' and 'things', so that all existing things are divided into two categories.[3] As a result, his characterization has a still more metaphysical flavor, logic not depending for its principles or notions upon "existing" things. Whatever the dichotomy notion is, it is thus not one of logic. Gödel therefore seems in error in stating that the set-theoretic paradoxes "are a very serious problem, not for mathematics, ..., but rather for logic and epistemology." By 'mathematics' here he means a set-theory as based on the iterative concept in which the paradoxes are not supposed to arise. Of course, Gödel is using 'logic' here in a broader sense than is now customary, a sense, be it said, that neither he nor anyone has ever made very clear, involving as it does unanalyzed intensions *sui generis* in such a way as possibly to lead to intensional antinomies (such, in the words of Wang, as "the concept of not applying to itself").

Gödel also speaks of his broad sense of logic in his paper on Russell[4], where he comments that "by analyzing the paradoxes to which Cantor's set theory has led, he [Russell] freed them from all mathematical technicalities, thus bringing to light the amazing fact that our logical intuitions (i.e. intuitions concerning such notions as: truth, concept, being, class, etc.) are self-contradictory." Nowadays we are bit more fastidious in our terminology and we have learned — or at least we *should* have — that the formulation of acceptable principles is more important than subservience to our naive and undisciplined "intuitions." Most of our intuitions are wrong most of the time, anyhow, and should not be heeded until taken through the furnace of critical humiliation. Fastidious philosophers are now inclined to classify concepts more finely than Gödel did: 'concept' itself is an epistemic term, 'being' a metaphysical one, 'truth' a semantical one, and 'class' one difficult to classify properly (see below). None of these are "logical" notions, strictly, and none of them need be self-contradictory when suitable principles concerning them are properly formulated. Gödel himself seems to recognize this point when he says, in a qualifying footnote (in Wang, p. 221), that "this observation by no means intends to deny the fact that *some* of the principles have been *formulated* quite satisfactorily, in particular all those that are used in the application of logic to the sciences including mathematics as ... "defined" in terms of the iterative concept of set." These principles are presumably precisely those of first-order quantification-theory.

It should be observed that the distinction between the "logical" and "mathematical" notions of 'set' is not very accurate. Wang himself, it will be recalled, speaks of a set as something "the members of which may be objects of any sort: plants, animals, photons, ... etc." If among the etc. entities here, the null set is included, we no longer have just the "mathematical," iterative view of set (as based just on the null set). Or we might have only non-sets among the *Urelemente*, in which case we do not have the mathematical view at all. Also the "logical" view, on the Gödel-Wang basis, makes fundamental use of 'the totality of all existing things' for its very statement. Is this totally itself a value for a variable in the formalism? Not in the original Zermelo view, although it may be so regarded in some reformulations of it. Also Wang's formulation of the dichotomy or "logical" concept of 'set' brings in "properties" also, so that it is presumably a property that is dichotomized into "those [things] which have the property and those things which do not." Strictly this view is a very different one from that of Gödel, and remains rather mysterious unless 'property' itself is well-defined or characterized, perhaps in terms of some 'ε' for the membership-relation of the system or even metatheoretically. It is to be feared that none of these views has been suitably spelled out with even the minimum of rigor now required in sophisticated philosophical study based on the new logic.

Philosophical objections to the notion of 'set' might all be countered with the casual comments that mathematicians "use the notion all the time," that, in the words of Zermelo himself, "set theory constitutes an indispensable component of the science of mathematics," and that sets are whatever the axioms tell us they are. These lines of defence, however, will not pass muster. Politicians, especially in academe, lie and cheat all the time; but to say this is not to defend it. The greatness and high utility of mathematics, especially of the theory of functions of a complex variable and of the various kinds of geometry, in no way depends upon its being either formulated or formulable in a set-theory. Also there are alternative approaches to mathematics that reject the use of sets altogether, e.g., the approach via some forms of algebra, constructivistic number theory, some forms of intuitionism, and others. The mathematical seriousness of some of these is sufficient to falsify outright any claim that set-theory is an "indispensable component" in mathematics. And, finally, the axioms of set-theory do not suffice to tell us what sets "are," but only how they behave relative to one another once we have them.

Wang himself is sensitive to these and related points. If we leave set-theory to the mathematicians, the irksome question remains, in his words, as to "the precise meaning ... [of the axioms] and why we accept such principles." This question "is incomparably more difficult" than the question as to "what, roughly speaking, the principles are by which we introduce the axioms." Wang comments that "it is my impression that Gödel proposes to answer ... the former question by phenomenological investigations." It would be interesting to have been told precisely how, the relevant phenomenology itself being in need of careful logical formulation in terms of an exact pragmatics.[5] Wang himself, however, is not friendly to the field of systematic pragmatics and pays no heed to the recent literature in it. As a result, he misconstrues (on pp. 383-384) its basic character and intent.

Let us observe now that, if we view mathematics either mereologically or homologically, as in V above, we have no need of hypothecating sets in any way, shape, or form. To be sure, much must then be made of individuals, including ordinal ones, and of Boolean sums of them. But this is a very different matter. One might, of course, insist dogmatically that an ordinal entity *is* somehow a set in disguise. Thus one might insist that the ordinal individual (x ord y) *is* merely the set $\{x,\{x,y\}\}$, or $\{\{x\}, \{x,y\}\}$ or $\{y,\{x,y\}\}$ or $\{\{y\},\{x,y\}\}$. To insist upon any such identification would be like insisting that any individual x *is* merely the set $\{x\}$. Such insistence would seem rather arbitrary, and would issue from the kind of set-theoretic arrogance bemoaned in VIII above. To all such talk one can now merely answer: *Je n'ai pas besoin de cette hypothèse.*

There can be no doubt, then, as a result of these various reflections, that all is not well with the notion of 'set' within mathematics, and that all is extremely ill with it outside. To be sure, the notion of 'individual' remains to be charaterized. Wang uses the notion without analysis, but then goes on to complain (p. 416) that "Quine and Goodman's nominalistic position [is] unclear; and this is because they have not explained ... what, according to them, are the individuals or nonclasses." By pairity of argument, Wang's initial characterization of 'set', quoted at the beginning, is then also "unclear." To say this is not to deny that we all should use 'individual' more carefully than we do, in each context of its use, and that probably no even minimally adequate characterization of it has ever been given.

The metaphysical attitude taken toward individuals in OM or HL leaves the matter open until we come to an explicative semantics of the object-language. An explicative semantics, it will be recalled[6], is always richer than the object-language of which it treats, in having a wider array of expressions available in terms of which the object-language primitives may be explicated. Where 'P' is the primitive for a relation of part-to-whole, we may well ask: what, or which, part-whole relation? The answer to this will be given, to some extent anyhow, in the semantics in which spatio-temporal considerations must be brought in if the 'P' is interpreted as applying to spatio-temporal entities, a theory of mental events if it is also to be applicable to such, and so on. The explicative semantics is open-ended in the sense that it need not be brought to a halt at any point, but it may be if we have gone as far as we can go on the basis of present knowledge or endeavor. The notion of 'individual', in a given context, is thus very much like any scientific or philosophical notion in being as "clear" in the given context as we are able to make it up to the time.

The important point here is that without *some* notion of individual, however inchoate or undeveloped, we cannot have any scientific or philosophical parlance at all. Nor can we have any "applied" mathematics. This point was well put years back by Josiah Royce. "For *logical* purposes," he wrote in 1914[7], "an Individual Object is one [— italics in the original —] that we *propose to regard at once as recognizable* or *identifiable* [not necessarily just in sense experience] *throughout some process of investigation, and as unique within the range of that investigation,*...." Further, "without objects conceived as unique individuals," he continued, "we can have *no Classes* [or sets]. Without classes we can ... define *no Relations*, without relations we can have *no Order. But to be reasonable is to conceive of order-systems, real or ideal. Therefore, we have an absolute logical need to conceive of individual objects as the elements of our [real or] ideal order-systems*. This postulate is the condition of any theoretical conception whatever. The further metaphysical aspects of the concept of an individual we may here ignore [for the moment]. *To conceive of individuals is a necessary presupposition of all orderly activity*" — and *a forteriori* of all orderly or systematic thought.

Royce's admirable statement here has been totally neglected by logicians and metaphysicians. Rephrase it ever so slightly and we have a fine and thouroughly defensible statement. By 'class' for the moment, let us understand the relata of a relation. We then have the view that "without objects ... we can have no relata, without relata we can have no relations, without relations we can have no order-systems." But, of course, "the metaphysical aspects of the concept of an individual" must be faced up to in the underlying logico-metaphysical order-system at the meta-level. Different features or aspects of the individuals are studied in more special "investigations." Logico-metaphysics abstracts from such study and, for example, need not *ab initio* deal with problems concerning how we *know* or are aware of individuals in perceptual, conceptual, or other experience. Also, at this point, let us take it as a truism that we do need order-systems for understanding mathematics, science, philosophy, and so on, and that "orderly activity" is in general a *good thing*.

What now is a *class*, as over and against a set? There seems to be no commonly accepted answer to this question. As good a usage as any, and perhaps better than most, would be gained by taking 'class' in essentially the sense of the simplified theory of types and leaving 'set' to the set-theorists (but allowing these latter, especially von Neumann and Bernays and their followers, also to use 'class' in a specialized mathematical way if they wish). If we do this, we could then use 'class' in a very exact way as a synonym for 'attributive relation' in either OM or HL. Such usage would accord well with custom and would provide for all extra-mathematical uses of the term. A relation R is an attributive relation relative to x, it will be recalled, if R is a dyadic relation such that

$$(x \mathrm{R} x \,.\, (y)(z)(y \mathrm{R} z \supset y = z)).$$

Thus an erstwhile one-place predicate 'Englishman', say, for the class of Englishmen, may be handled here as an attributive relation 'is the same Englishman as'.

Another suggestion is that we use 'class' wholly in the sense of a *virtual* class. Outside of mathematics and its applications, we can usually get along with only virtual classes. It would be more correct to speak here only of the virtual-class *notation*, virtual classes being mere fictions. It is this that is useful and helps us to compensate for the lack of a theory of "real" classes as values for variables in a given formalism.[8]

Another item to note concerning sets is the popular identification of them with Platonic ideas. It has been urged in VII above, it will be recalled, that this is erroneous and rests on an inadequate reading of the Platonic *corpus*. Father Joseph Owens, the eminent Aristotle scholar, in a recent review[9], laments "the use of modern terminology to explain Aristotelian notions, for instance to present the Aristotelian universal as a "class" or "set" ... in a sort of "agreed metalanguage"." This lament may be made *a fortiori* for Platonic "ideas," of *eidé*, particularly in view of the widespread use of 'platonic' as almost synonymous with 'set-theoretic'.

An interesting methodological argument against the admission of an ontology of sets is as follows. One's ontology, it may be thought, should be all of a piece. The fundamental ontology of mathematics should not differ in kind from that required in physics and the other sciences. It has been urged, in X above, that we live in a structured-world-of-cardinal-and-ordinal-individuals-in-indissoluble-interrelationship-with-one-another. Such a general ontology suffices for the sciences and for mathematics alike. Each science will pick out special features for special study, as we have already noted, but the basic logico-metaphysical aspects of the individuals remain the same. These aspects are characterized in OM or HL, at least in a tentative, and exploratory first approximation.

Another argument against sets is the celebrated principle of parsimony, that *frustra per plura quod per pauciora faciendum potest*. The methods of OM and HL show how this principle may be satisfied without loss: simply do away with all sets and use mereology or homology instead, bringing in such additional transfinite entities and axioms governing them as might be needed, for example, for mathematical theories of the infinitely large or the infinitely small. And bring in suitable non-logical constants to handle such scientific notions as may be needed for particular extra-mathematical purposes.

It is also worth observing that any set-theoretic rendition of mathematics may be likened to a huge iceberg, the greater portion by far floating beneath the surface to support the visible mathematics above. The fact is that only a very small portion of set-theory is actually needed for the purpose of providing for mathematics in the great classical sense. This seems rather odd: too inflated a price to pay, both as to ontology and axiomatic structure, for the pure mathematical gold to be gained. Let us go after the gold more directly.

One might think that a similar objection could be made against OM or HL. But no. It should be recalled that the axioms of existence of these latter are rather restricted, and postulate the existence of just such entities as are needed for specific mathematical or other purposes. Of course, this method is open-ended, and one can add stronger existence assumptions when needed. It might be thought also that the type-theoretic rendition, used in the further development of OM or HL, would restrict mathematics to a good deal less than is contained in set-theory. This, however, seems not the case. Apparently nothing of mathematical significance depends on the use of set- rather than type-theory that cannot be provided for in this open-ended approach.

It has been remarked above that set-theory, when formulated in terms of the iterative notion of set with the empty set as the only *Urelement*, cuts mathematics off from the world. This is not the case if *Urelemente* other than the empty set are admitted. But if the latter are admitted, and a calculus of individuals is introduced to help characterize them, we can eliminate the sets altogether and still do mathematics. And we can also do logico-linguistics. Without individuals, or *Urelemente* other than the empty set, we could not give logical forms for natural sentences in any direct way. Of course we might force set-theoretic notions upon them, but this would be *ad hoc* and at best Pickwickian. It is especially interesting to note that no set-theoretic notions whatsoever are made use of in Zellig Harris's recent *A Grammar of English on Mathematical Principles*,[10] a *locus classicus* where they might be thought needed fundamentally.

One point made by Wang seems to be in favor of such methods as those of OM and HL, and in general helps to substantiate some of the comments made above. "In my opinion," he writes (p. 422), "a comparative study of a calculus of individuals (with infinity assumed) and the ordinary class theories might lead to illuminating distinctions. Such an investigation is, I think, one direction which could perhaps make the position of rejecting everything except individuals significantly fruitful."

To summarize. Several arguments against admitting an ontology of sets have been made in the foregoing. Set-theorists are in general unhappy about the subject, and so are many constructivists, intuitionists, and applied mathematicians. Too many dubious assumptions and entities are needed for it. Mathematics occupies merely the tip of the set-theoretic iceberg. One eminent mathematician, at the International Congress for Logic, Methodology, and Philosophy of Science at Bucharest in 1975, is often quoted as having said that "set-theory has been an utter failure in mathematics." If so, there is little reason to think that it should be a success when applied elsewhere, where the use of virtual classes instead has often proved helpful. Simpler methods for handling mathematics are now available — 'simpler' in the sense of involving a clearer ontology and less dubious assumptions. Such methods suffice also for logico-linguistics. The rejection of sets is not anti-platonic, and anyhow we should not read set-theoretic terminology into the great philosophical writings of the past. For a unified logico-metaphysics, the only entities that should be needed are those that are needed *überhaupt* in science and elsewhere; the admission of sets is at best an *ad hoc* device within mathematics alone. The principle of parsimony remains a valuable methodological guide. Such counterclaims as "Nature loves a plethora and abhors a vacuum" miss the point entirely; the principle of parsimony commends precisely what nature contains, plethora and vacua alike. Set-theory has had the unfortunate effect of cutting mathematics off from the world and from the helpful role it has had traditionally, and indeed should have, in other domains of inquiry.

CHAPTER ELEVEN

On the mainstream of science and Suppes's pluralism

"Ne depugnes in alieno negatio."

In a provocative recent paper[1], Patrick Suppes has put forward some challenging comments in response to a critique of his views.[2] He centers his comments around (1) the problem of a unified language for science, (2) the role of set-theory in science, (3) the language of events in science and ordinary talk, and (4) problems of intensionality. As usual in such replies to criticism, only a few of the substantive points raised in the original critique are considered; most of them, as well as the relevant literature concerning them, are passed over in silence. Let us not do likewise now, but instead consider Suppes's reply item by item, at about the same level of both generality and technicality as in his discussion.

The most basic item seems to be (2), that of the role of set-theory in science. "One of Martin's points," he writes, "is that my use of a set-theoretical framework is mistaken, for such an apparatus is not needed scientifically. He proposes instead to use various philosophical variants, such as an axiomatized Boolean algebra or a language of part-whole as exemplified by mereology. I think he is flatly and unequivocally wrong. The idiosyncratic languages he talks about are of interest in philosophy but for very special reasons, and they cut off philosophical discourse about causality from the mainstream of scientific talk." Unfortunately for Suppes's argument, the first two statements quoted here, as well as the last, are not correct. His use of a set-theoretic framework was not said to be "mistaken," and the event-theoretic framework advocated consists of a great deal more than an axiomatized Boolean algebra or mereology.[3] Nor does the use of the event-theoretic framework "cut off philosophical discourse" from the scientific mainstream. On the contrary, it seeks to give this latter a firmer and more unifed logical foundation than it would otherwise have.

With this last observation, we come to the key point of disagreement. Suppes is interested, in the paper under review, in the scientific mainstream at the surface level, but not in the logical foundations needed as a basis for justifying it, criticizing and improving it where appropriate, integrating it where possible, and so on. No, Suppes accepts the *status quo* of the mainstream and of set-theory without qualms. (When it gets into trouble, all he can do is say: woops, sorry!) And the uncritical use of set-theory goes hand in hand with this acceptance. Suppes in effect defends the whole of set-theory whereas actually it is only a small fragment that he makes use of in his technical work. His position would be a stronger one if he were to argue for the acceptability of just the set-theoretic principles he actually uses for given causal-theoretic, learning-theoretic, and other purposes.

In the Appendix to his paper, Suppes puts forward two general theorems in learning-theory. He remarks that "none of the more complicated set-theoretical machinery I consider [in discussing these theorems] is duplicated in any way by Martin.... The kind of language moves that Martin proposes would lead to further isolation of philosophic talk about these matters, an isolation that has already been too noticeable in the literature on confirmation and the foundations of induction." It is true that the machinery needed for Suppes's theorems is not duplicated by me in the paper he is discussing, but this is not to say that it cannot be. All one needs to do to show that it can is to pick the proper framework — say OM or HL or even weaker theories[4] — and get to work on the details. It is thus by no means the case that the use of these frameworks leads to the kind of isolation Suppes laments. It is curious that he laments this isolation but yet espouses scientific pluralism, this latter being virtually tantamount to an extreme form of scientific isolation, as we shall observe in a moment.

Suppes is "quite willing to accept that with enough effort, all of the standard machinery characteristic of modern probability theory, not to speak of statistical theory, could be built up in one of Martin's idiosyncratic frameworks. But this would seem to me a terrible waste of time, and at the same time would isolate the developments from the large and interesting literature in science, mathematics, and statistics on these matters." The two contentions here are not only incorrect but run counter to one another, it should be noted. The EM and allied frameworks are unifying ones, helping to bridge the gaps — indeed the chasms — between and among mathematics, the sciences, natural language, exact philosophy, and even metaphysics. And why the terrible waste of time? Not much time would be required, once one got the hang of it, no more than is needed in going from ordinary set-theory to its applications.

Suppes is one of our most distinguished philosophers of science and deserves high praise for having called attention over the years to the importance of the axiomatic method, particularly in set-theory and in empirical applications. The day of set-theory, however, seems to be waning. More and more the view seems to be that it has proved an "utter failure" — recall X above — in providing adequate foundations for mathematics. Suppes can easily — and indeed should — defend the axiomatic method without hitching it as intimately as he does to set-theory. As already noted, much less than a full set-theory can be made to suffice for many of his purposes. Also, many alternatives are on the market clamoring for attention — category-theory, intuitionism in at least five or six versions and guises, various forms of constructive type-theory and of constructivism in general, and so on and on. In any partiular application of mathematics to empirical science, an interesting problem for Suppes is for him to "keep tabs on his tools" at each step, and to note at the end how much or how little of the far reaches of set-theory he actually has needed.

Let us turn now to (1) pluralism and the problem of a unified language for science in general and for scientific causal talk in particular. For pluralism to have any bite is for it to have an open attitude toward the multiplicity of alternatives, toward competing theories, toward theories incompatible with some favored one, and so on. Curiously, Suppes's attitude, in the paper under review, is not in accord with this key tenet. Theories other than his favored one or ones he presumably dubs 'idiosyncratic'. Just what he intends by this word he does not say, nor does he give us a list of theories to which it is to apply. Just which of the mathematical alternatives mentioned above are idiosyncratic, according to him, and just why? Also he seems to condemn by innuendo languages that "are of interest to philosophy but for very special reasons." But, surely now, all languages that are of interest to philosophy are so for special reasons, some perhaps for *very* special reasons. And these may be for *good* reasons, furthermore — perhaps even *very* good reasons. Pluralism should welcome these languages and seek to assess their relevance for given portions of science and mathematics, and above all reflect upon the cogency of the special or very special reasons put forward for their advocacy. Suppes fails to do this for any of the languages suggested, dismissing them without even a hearing. His remarks about pluralism here seem totally at odds with the content of his valuable and more serious discussion of the subject in "The Plurality of Science."[5]

This is not the occasion to review at any depth the thesis of the unity of science of the Viennese positivists, but merely to call attention to some of its leading features that are still very much alive and kicking. As Max Black noted years back[6], one feature of the view is that "speculative philosophy must be transformed into a new methodology, the analysis of linguistic forms," and that "the chief concern is to consolidate the achievements of scientific discovery by analyzing the limitations and essential structure of the language in which all knowledge must be expressed." In 1934 the limitations were as significant as the essential structure. During the fifty years that have elapsed since this passage was written, however, an enormous develoment has taken place in both structural linguistics and in the logical study of natural language. The new science of logico-linguistics is developing in which more and more of "essential structure" can be accomodated and in which the "limitations" no longer loom large. The thesis of the unity of scientific *language* has much greater cogency today than it has ever had. This again is an important fact totally neglected by Suppes.

Let us turn now to (3) concerning the language needed to talk about events, both scientifically and in ordinary language. Here again Suppes takes the "pluralistic view that it is probably not possible to give a tightly unified account of these many different uses." We should distinguish here giving a unified account from employing a *unified ontology* in terms of which such an account, or indeed any account, could be given. It is this latter that Suppes thinks somehow suspect. On the event-logical theory, events themselves are taken as fundamental object, whereas for Suppes events are handled as "sets of outcomes." But the logical character of these latter is not fully characterized — a careful theoretical foundation for them being needed fundamentally.

Suppes regards his view as the "standard" one among probability-theorists and comments that this view is objected to in the critique. But this is not the case. The objection put forward is that Suppes provides no appropriate foundation for his view. (Also he seems not to have read the passages (pp. 201-2) in the original critique showing how his "sets of outcomes" may be handled within the event-theoretic framework.) At the same time, he admits that (p. 367) "reforming or changing the standard concepts of probability ... is a useful matter, but it is not even useful when it is idiosyncratic in the way that Martin's discussion is." Here again, it would be interesting to be told in just what way introducing event-variables, and distinguishing sharply events as such from "kinds" or sets of outcomes (or "attributive relations" in the parlance of OM or HL), is idiosyncratic other than as not being explicitly the method of Kolmogoroff, the one favored by Suppes.

Again, Suppes urges that "adopting Martin's recommendations would isolate philosophical discussion of causality" and of probability, and he thinks that "a consequence of that isolation would be consideration only of the most elementary points about causality." On the contrary, Suppes's view here is the isolationist one, limiting probability talk to within the narrow confines in which sets of outcomes are of interest. And had Suppes read the relevant passage in which it is shown how these latter may be handled within the event-theoretic framwork — and *a fortiori* within OM or HL — he would have seen how all manner of non-elementary points about causality may be discussed.

Suppes goes on to admit that "in a certain sense, the notion of event as having definite physical properties [a notion fully provided for in the EM-framework, it should be noted] is treated in a rather cavalier fashion in standard probability theory." Quite, and this is one of the points of the original critique. "When we are concerned with more specific philosophical questions or more specific physical questions," he continues, "as for example a discussion of how one should use the concept of event in relativistic physics, we may want to say a good deal more [than is provided by Kolmogoroff]. I certainly want to admit that such extensions are proper but I also want to make clear that I think the direction Martin takes the discussion is mistaken." It would be interesting to be told precisely how and why. The only hint Suppes provides is to the effect that he thinks it "mistaken to tie down the concept of event [in general] by more detailed assumptions, as for example the kind that are easily suggested by physical theories." Quite, and no such tying down is made or in any way demanded in the event-logic framework. On the contrary, ample means are provided for handling all manner of social and economic phenomena, and the need for recognizing even a realm of mental events in some sense is particularly emphasized. Suppes seems to show so little understanding of key items in the original critique and of its background theory, that it seems a bit cavalier of him to be so sure — and to iterate so many times — that the direction it takes is "mistaken" and "idiosyncratic."

Let us turn now to (4), concerned with intensionality. Suppes agrees that "the standard set-theoretical framework I [Suppes] adopt does not give a fully satisfactory treatment of intensional matters." Quite, as was emphasized in the original critique. Curiously he thinks also "that a full-blown apparatus to handle these matters in completely explicit fashion will be another step toward the isolation I have already spoken of." At the same time he urges "a move from set-theoretical to [so-called] procedural semantics," unaware apparently that the logical foundations of this latter can be handled in terms of a systematic pragmatics incorporating the event-theoretic framework.

This is not the occasion to attempt to provide a formulation for Suppes's procedural semantics[7] in the full detail required — an attempt which he seems not even to have made himself. Instead, let us glance at it just enough to see how this could be done within the framework being advocated here. The key items of Suppes's theory are that "the meaning of a word, phrase, or utterance is a procedure, or collection of procedures," that "the meaning of a word, phrase, or utterance is private for each individual," that successful communication depends upon the possibility that some spoken (or written) words, phrases, or utterances may be identified as "being the same," or being appropriately "congruent," by speaker and listener, and so on. A procedure is thus presumably a kind of linguistic act involving speaker and listener, so that, for example,

(1) '$p\ \mathrm{Utt}_L\ a,q,e$'

may express that person *p utters* the expression *a of* the given language *L to* person *q on* or during the occasion *e*. Utt here is a certain fundamental relation of systematic pragmatics. The many kinds of relations of "being the same," congruence-relations, may be handled, each in its own way, in terms of analogous locutions. Many such relations have been suggested in the relevant literature, beginning with *Toward a Systematic Pragmatics*. (Recall III above.)

Suppes contends that "classical set-theoretic referential semantics can be obtained, where appropriate, by abstraction from procedural semantics." The "abstraction" required here is of course very complex. A full characterization of it would involve several stages of methods of generalizing from sentences of the form (1) and allied forms, to sentences for some or all words or phrases (inscriptions) of similiar kind typographically, for some types of occasions *e*, and for suitable *p*'s and *q*'s in given idiolects, social groups, and so on. Some of these abstractions lead us into psycho- and socio-linguistics, and some of them into standard semantics in which the designata and denotata of the various words are thought to be fixed once and for all. Abstractions of this last type are of course synchronic and thus highly artificial, and hold only for the language at an instant, so to speak. More important abstractions must be diachronic, so that a full theory of events, changes, procedure, and the like, must be built into the interlinguistic, comparative metalanguage needed for formulating the theory of procedures.

Suppes contrasts procedural semantics with set-theoretical semantics rather too sharply, not realizing that the former itself, once properly formulated, may be given a semantics either set-theoretically or otherwise.[8] "It is a surprising and important fact," he writes, "that so much of the language of mathematics and science can be analyzed at a nearly satisfactory formal level by set-theoretic semantics, but the psychology of users is barely touched by this analysis." Of course not, this is not its intent. The psychology of users is to be handled in a pragmatical metalanguage containing the requisite vocabulary, which metalanguage in turn can be given a semantics set-theoretically (or otherwise more simply).[9] It is a grave methodological blunder to foist on semantics the whole burden of characterizing the vocabulary describing the psychology of the user, including "ordinary verbs of perception," "the many nouns and adjectives describing emotional states," and so on.

Suppes comments that "in the case of ordinary language the set-theoretical fare does not offer minimal nutritional content. True, we can have a certain amount of logic, but the gruel of quantifiers and connectives is too thin. Not only are the four-letter words missing (in the case of English [the Anglo-Saxon monosyllables, no doubt]), but also the much more important high-frequency two-letter words: *to, in, of, on, at, by* and *as*. The set-theoretical semantics of any of these seven two-letter words — or their functional analogues in other languages — is far from being satisfactorily characterized." Quite. But note the curious confusion here between what a semantics is supposed to do with what can be done in logico-linguistics with the introduction of suitable non-logical primitives for *prepositional* relations.[10] Suppes seems unaware of these latter, and yet their use would seem quite essential for any rigorous presentation of procedural semantics. Of course they cannot be introduced just in terms of the 'ε' of set-theory, the connectives and quantifiers, and semantical notions. No one would presume this any more than he would suppose he could get relativity-theory or quantum-physics out of such meagre fare. The prepositional relations are very exact ones, however, and may be characterized by suitable meaning-postulates. And of course there are many more than just Suppes's seven — about 75 basic ones in English.[11]

The whole field of logico-linguistics has been developing in recent years in such a way as to provide logical forms for all manner of natural sentences concerning perception and emotional states. At any stage of development, the theory of logical forms can be given a semantics in whatever fashion is desired. Suppes seems unaware that the very methods of procedural semantics — or their event-theoretic counterpart — are helping to pave the way for an increasingly more rigorous and more extensive theory of logical form for natural sentences. Whether set-theory is needed anywhere along the line, or not, is a moot point. It is interesting to observe again that, in his early work, Zellig Harris made much of set-theoretic notions, but has found over the years that they are not needed. In his recent, monumental *A Grammar of English on Mathematical Principles*, no trace of them is to be found. (On set-theory, recall X above.)

In the original critique of Suppes it was suggested that causal talk, both in science and in ordinary language, be handled intensionally. Suppes thinks that the intensional treatment "is not important in the framework of developing a causal theory for scientific purposes.... On the other hand," he continues, "I think Martin is right in insisting that such intensional matters can be important in the sensitive analysis of causal concepts as they are referred to in ordinary talk." Well, Suppes is no doubt right in thinking that causality need not *always* be handled intensionally. The view taken here is that the non-intensional handling of causality is merely as a special case of the intensional one. Let

'e_1,a Cause e_2,b',

express, as previously, that event e_1 as taken *under* the one-place predicate-description a is a cause of e_2 as taken under b, 'cause' here being construed in the tense of timelessness. Then the non-intensional form

'e_1 Cause e_2' may abbreviate '(Ea)(Eb)(e_1,a Cause e_2,b)'.

Similar definitions could be given construing the 'e's here as ranging over "sets of outcomes" rather than over procedures or events as such. Thus, it would seem, there need be no further quarrel as regards causality either in or out of science.

Suppes closes his discussion by commenting that he hopes "to help keep future efforts at analysis from being too much diverted from the mainstream of science to idiosyncratic philosophical bayous." The attitude suggested by this is surely not in accord with an open-minded pluralism, as already noted. Nor is it tolerant of alternatives, being directly opposed to Carnap's much-espoused Principle of Tolerance. Also it places in the bayous views concerning events stemming from Peirce, Whitehead, Russell, Carnap himself, Reichenbach, Uriel Weinreich and other linguists, Donald Davidson, and others — views that surely should not be labelled 'idiosyncratic'. Also it places set-theory at the very source of the mainstream, thus blocking the road to inquiry concerning more constructive types of mathematics that occupy so much attention on the part of contemporary mathematical logicians.

CHAPTER TWELVE

On the semiotics of mathematical praxis

> "Nemo mathematicus genium indemnatus habebit."

"At the turn of the century, the Swiss historian Jacob Burckhardt, who, unlike most historians, was fond of guessing the future, once confided to his friend Friedrich Nietzsche the prediction that the Twentieth Century would be the "age of oversimplification"," Gian-Carlo Rota tells us in his Introduction to *The Mathematical Experience*.[1] The prediction itself is no doubt an oversimplification, as are the comments concerning philosophy that Rota goes on to offer. "Philosophers have [after WW I] proposed daring reductions of the complexity of existence to the mechanics of elastic billiard balls; others, more sophisticated, have held that life is language, and that language is in turn nothing but strings of marble-like units held together by the catchy connectives of Fregean logic." To be sure, mechanism in one form or another is one of the enduring metaphysical views, modern forms of which are often subtle in the extreme, "elastic billiard balls" having been left far behind. And no serious philosopher, surely, has ever held that "life is language" in any reasonable sense of 'is'. Presumably sentences and texts are "strings of marble-like units" whose inner structure and semantics remain invariant when being used in normal ways. Does Rota wish them to melt away or to change substantially during their use? And why are the "catchy connectives of Fregean logic" in any way objectionable? No more or less so, surely, than numbers, differential equations, Lebesgue integrals, and so on, all of which manage to catch a good deal in their stride. Moreover, sentences concerning them contain the Fregean connectives in one way or another in their deep structures.

Rota comments that Russell and Gentzen "command that mathematics become logic," as though this is in some way incompatible with the fact that "some mathematicians are more endowed with the talent of drawing pictures, others with that of juggling symbols and yet others with the ability of picking the flaw in an argument." All such activity is to be welcomed and to be valued by its results. "Juggling" symbols is by no means the least valuable, as much of the history of mathematics attests. The juggling must, however, be done with great care in order to gain a result of any interest or value. Russell's *fiat mathematica logica*, if such it be, surely helped pave the way for something valuable, if only as one chapter in the history of mathematical rigor in juggling. Drawing mathematical pictures has its rigor, too, let us not forget, as does picking flaws in arguments. Special gifts are required for each.

Rota seems to object to the contention that mathematics consists mainly in "proving theorems" on the grounds that a mathematician's work rather "is mostly a tangle of guesswork, analogy, wishful thinking and frustration, and proof, far from being the core of discovery, is more often than not a way of making sure that our minds are not playing tricks." Here there seems to be confusion between means and ends, between discovery and achievement, between discovering "facts" or theorems and proving them rigorously. Both types of activity are surely welcome and necessary. The wedge Rota forces here between them seems to be a misleading oversimplification. "To master mathematics is to master an intangible view, it is to acquire the skill of the virtuoso who cannot pin his performance on criteria." *He* cannot perhaps, but we who listen can, provided we are properly trained listeners. The shallow, superficial, showy virtuoso is easily separated from the self-effacing master interpreter, even though it is very difficult to spell out in full detail the criteria for the latter.

Rota's kind of oversimplification permeates much of the discussion by Davis and Hersh also. Let us examine, in what follows, the most important of these as far as concerns philosophical and logico-methodological matters. The authors seem unaware of the purport of logic, and also of certain newer developments, that would enable them to state more carefully and forcefully some of the worthwhile points they make.

Let us start with some comments on symbolization. The authors state (p. 123) "the principal functions of a symbol in mathematics are to designate with precision and clarity and to abbreviate." They quote Whitehead approvingly, from a famous passage, that "by relieving the brain of all unnecessary work, a good notation sets it free to concentrate on more advanced problems, and, in effect, to increase the mental power of the race." This cannot be done without at least some degree of precision, without requiring (p. 123) "that the meaning of each symbol or each symbol string be razor sharp and unambigous." How is this achieved? The authors speak of "designating with precision and clarity," not realizing that they are thereby invoking semantical rules of the most basic kind, namely Rules of Designation. (Let us use capital letters here in accord with the dignity and high importance of what is under discussion.) Such rules state precisely what each (primitive) symbol (proper name or predicate) designates, and that it designates at most one entity.[2] There are, to be sure, other types of semantical rules, but the authors, in their use of 'designating', have hit upon the most basic by far. And, similarly, the authors correctly realize the importance of abbreviations, of giving definienda for definientia, the former usually being shorter and more concise versions of the latter. "In point of fact [p. 124], without the process of abbreviation, mathematical discourse is hardly possible." Nor is any other kind of discourse. Nor, for that matter, is any kind of discourse possible without some fixed vocabulary subject to Rules of Designation — ambiguity, vagueness, and the like, then suitably to be handled within the wider context of the study of usage, communication, and the like.

What now do we do with symbols? "We act [p. 125] on mathematical [and other] symbols in two different ways: we calculate with them, and we interpret them." To calculate is to make use of "a standardized set of agreements" enabling us to convert one or more strings of symbols into others, in other words, to use the Rules of Inference provided. "Interpreting a symbol," however, the authors tell us, "is to associate it with some concept or mental image, to assimilate it to human consciousness ... [and] the rules for interpretation cannot be any more precise than the communication of ideas among humans." Some rules of "interpretation" are semantical, and some are epistemic or pragmatic. The former fix the "razor sharp and unambiguous" features of symbols; the latter describe or characterize features of a very different kind. It is a pity the authors do not distinguish clearly these two types of rules, giving unto each the uses and merits thereunto appertaining.

The authors go on to note that the "process of representing mathematical ideas in symbolic form always entails an alteration in the ideas: a gain in precision and a loss in fidelity or applicability to the problem of origin." Let us consider the finite cardinals and the problem of counting. What alteration or loss is undergone when we symbolize seven as '7'? It would seem rather that there is alteration only in the direction of gaining more razor-sharpness and unambiguity. In place of something we know not what we now gain something subject to the exact rulers of arithmetic. And what loss is undergone if we now say that 7 is the number correctly to be assigned to some class or set such as $\{x_o,y_o,z_o,w_o,v_o,t_o,u_o\}$, where x_o, y_o, and so on, are distinct objects in the domain of objects under consideration? On the contrary, there would seem to be gain here too. And as goes the numerals for finite cardinals, so goes the whole regalia of other mathematical symbols. The authors' point here, with equal justice, could be made about the words of a natural language, with the absurd follow-up that language is in no way faithful or applicable "to the problem of its origin," i.e., with what is intended to be said, say, in any given sentence. Language is not identical with the sentence or problem, of course, but we express this latter in the former, not always capturing every little nuance, of course. Wherein we fail, we go to try to do better or to find some other means of doing so.

In discussing "the hierarchy of intellectual values," our authors comment (p. 305) that "the level of intellectual acclaim [given to the solution of some problems] is proportional to the apparent complexity of the abstract symbolization.... Scientific education is often directed not to the solution of specific problems, but toward carrying on a discourse at the highest possible intellectual level." What is this level? And how do we attain it? It would have been good to spell out in detail of just what this highest discourse consists. The Zermelo-Fraenkel-Skolem system, the Russell simplified theory of types, and many another system, might be thought to come fairly close, were it not for their internal complications, their ontic extravagance, and the conceptual obscurity of their basic notions. It is not the inner content of such theories that is to be extolled, but rather the attempt to formulate them, as deductive systems with explicitly stated axioms, rules, and definitions, with about as much razor-sharp clarity and unambiguousness as we mortals have ever been able to achieve in any field of inquiry. Such systems should be looked upon as paradigms of methodology, not as paradigms of theory. The theories may be quite unacceptable, but the way in which the attempt has been made to formulate them is admirable and beyond cavil. The systematicity thereby attained is pure music to the intellect, and of the highest beauty, and mankind will probably never succeed in doing better methodology-wise. All science aspires to the condition of this music, to borrow a phrase from Walter Pater, and is never content until it attains it.

We should not make too much of the "complexity of the abstract symbolization." The symbolization is easy once one has thought out what it is that is to be symbolized. The most difficult material to think about clearly is usually the simplest in some sense. Think of the long, hard road that has had to be traversed — a 2000 (or so) year journey — to become clear about the truth-functional connectives, the quantifiers, and identity, to say nothing of whatever one wishes to add to this list. Only after some modicum of clarity about these was attained have we been able to characterize the systematicity of mathematics, either in part or as a whole, with any even approximative exactitude. Without such systematicity, any axiom-set is something of a joke: Euclid's, Hilbert's, Zermelo's, Russell's, E.V. Huntington's, Tarski's, and so on and on.

The authors make no effort to view systematic and epistemic matters in proper relation to each other. "The definition-theorem-proof approach to mathematics [p. 306] has become almost the sole paradigm of mathematical exposition and advanced instruction," they observe disparagingly. "Of course, this is not the way mathematics is created, propagated, or even understood. The logical analysis of mathematics, which reduces a proof to an (in principle) mechanizable procedure, is a hypothetical possiblity, which is never realized in full. Mathematics is a human activity, and the formal-logical account of ... [it] is only a fiction; mathematics itself is to be found in the actual practice of mathematicians." But, surely now, no one has ever seriously maintained that "the definition-theorem-proof approach" is *the* one and only way mathematics is ever "discovered, propagated, or even understood." It may play a role in discovery and its propagation and be helpful for understanding. It is there as a backdrop, so to speak, in one form or another, in mathematical practice. It would have been interesting for the authors to spell out exactly and in detail precisely how. Here is part of the real subject-matter of their book, which unfortunately they never get to. Of course mathematiz*ing* — doing mathematics — is a human activity consisting of an enormous variety of types of acts, mostly mental but not completely so, each type having its proper role and importance. To give a "natural history" of these types, of how they interrelate with each other, and of how they may relate to axiom-formulation, definition-formulation, theorem-statement, and to the carrying out of proofs, would be a valuable contribution to the study of mathematical praxis.

Surely the statement of theorems, algorithms, and the like, should be regarded as a part of mathematical praxis; but even so, we should distinguish proofs as such from acts of actual and purported prov*ing*. In some acts, or a succession of such, correct proofs are carried out or "given." These acts do not constitute the proof but are associated with it in a certain way.[3] The authors state apodictically that the "logical analysis of mathematics ... is never realized in full" and in fact "is only a fiction." If so, the appropriate response should be: what a fiction! let us have more like it! Even so, the use of 'fiction' here seems ill-chosen. 'Idealization' would be better, and we all know how hard it is to formulate ideals, and to stick to them once formulated. Just because they are difficult to attain is no reason whatsoever why we should abandon their pursuit. The authors, in some of their moods, seem to dislike this "formal-logical" ideal, as a result no doubt of their failure to understand it very deeply, its intent as well as its limitations.

The key chapter of philosophic interest in this book is the one entitled "From Certainty to Fallibility." Throughout this chapter we find the same confusion between theory and practice and between theory of theory and theory of practice. From the standpoint of logical analysis, we are told (p. 344), "a problem of principal concern to the mathematician becomes totally invisible. This is the problem of giving a philosophical account of the actual development of mathematics, of preformal mathematics, ..., including an examination of how this preformal mathematics relates to and is affected by formalization." This problem is not "invisible" to the methods of logical analysis unless one wears blinders. Enter the hero, Imre Lakatos, who will set the matter straight by reiterating the Popperian commonplace observation that "mathematics too [in the sense of mathematiz*ing*], like the natural sciences, is fallible, not indubitable: it too grows by the criticism and correction of theories which are never entirely free of ambiguity or the possibility of error or oversight." And so on and on, as though no one had ever said or thought such things before in the entire history of the subject. The authors seem ill-advised to extol Lakatos's *Proofs and Refutations*[4] in the way in which they do.

Lakatos contends that formalism, that school "which tends to identify mathematics with its formal axiomatic abstraction and the philosophy of mathematics with metamathematics ... disconnects the history of mathematics from the philosophy of mathematics.... Formalism denies the status of mathematics to most of what has commonly been understood to be mathematics, and can say nothing about its growth." All of these contentions are in error. Formalism, in the sense of Hilbert, did not "identify" mathematics with its "formal axiomatic abstraction." Formalism does, of course, make fundamental use of formal axiomatic systems, subjecting them to detailed analysis as to the consistency of their axioms, their independence, completeness, and so on. This is not to deny the importance of all the pre-formal hard work leading up to the axiomatic system, however. This hard work is quite indispensable, but it is not the purport of metamathematics to describe it. If one steps out onto the stage to give a piano recital, he should not be expected then and there simultaneously to give a violin recital also. Nor should he be accused of "disconnecting" the latter from the former or of denying its proper status. Lakatos's blunder here is precisely of such a character. To instigate a programme of metamathematical investigations is not to do history, but it is not to rule it out either.

Lakatos contends that "the formalist philosophy of mathematics ... is the latest link in the long chain of dogmatist philosophies of mathematics. For more than 2000 years there has been an argument between dogmatics and skeptics. In this great debate, mathematics has been the proud fortress of dogmatism.... A challenge is now overdue." This contention is based on a superficial reading of history and tends to confuse mathematics itself with popular, almost always erroneous, opinions about it. If mathematics has in fact been something of a "proud fortress" during these 2000 years, there may well be good reasons quite independent of any "dogmatism." What are these reasons? Have they ever been spelled out? Probably not. Any such spelling out would have to take account of the *grandeur* of the subject, of the deep inner certitude of its findings once understood, of the almost universal consensus concerning them, of the paradigmatic role and centrality it plays in all of our knowledge, of its enormous success in applications, of its unified character, and so on and on. It is a pity that Lakatos and our authors did not attempt to articulate these reasons, or to characterize the indisputable greatness of mathematics independent of "dogmatism," on the one hand, and skepticism, on the other, in such wise as to call for no "challenge" beyond calling for greater factuality and refinement of detail.

The authors think that derivations or proofs carried out in terms of "first-order logic" are "purely hypothetical activities." "The real situation," they write (p. 354), "is this. On the one hand, we have real mathematics, with proofs that are established by 'consensus of the qualified.' ... Even to the qualified there are normally differences as to whether a real proof (i.e., one that is actually spoken or written down) is complete or correct. These doubts are resolved by communication and explanation, never by transcribing the proof into first-order predicate-calculus." This is, of course, a dogmatic contention and the presence of 'never' surely renders their contention too strong. And even so, the contention is misguided and beside the point. If one can "communicate and explain" what one sees without a telescope, say, all well and good. The discovery and use of telescopes has led to seeing all manner of new things and has widened the domain of the scientifically visible. And so with the predicate-calculus. It has helped to make our notion of "proof" more exact, to recognize different kinds of proofs where it was previously thought that there is only one, to see that what were previously called 'proofs' were not so at all, and so on. But "fallibilism" still holds with regard to proofs carried out in a logical formal way. We or the computer may well be in error at some point. The more logical tools are self-consciously used in mathematical praxis, the more the likelihood of such errors is reduced. We should distinguish here, as the authors do not, minor human errors from strictly logical or mathematical ones. The former are always with us; the latter we try to correct by continual improvement of our logical theories as well as of our mathematical ones.

"Spokesmen for formalism," we are told (p. 356), "... never explain in just what sense formal systems are a model of mathematics. Is it in the normative sense — mathematics *should* be like a formal system? Or is it in the descriptive sense — mathematics *is* like a formal system." The answer is: neither. Mathematics *can be shown* to be "like a formal system." To say this, of course, is not to say that formal systems give us "ultimate foundations." Nothing seems ultimate in human knowledge. But formal systems do give us foundations, nonetheless, and the search for satisfactory, and even "convincing," foundations will go on and on, if only for the dignity of the mind itself, in Hilbert's famous phrase.

The authors at one point (pp. 358-9) seem to recognize the inadequacy of much of what they have been saying. They comment that, after all, "mathematics is one single thing. The Platonist, formalist, and constructivist views of it are believed because each corresponds to a certain view of it, a view from a certain angle, or an examination with a particular instrument of observation." "Our problem," then, they continue, "is to find an understanding of the thing itself, *to fit together the partial views* [italics added] — each of which is wrong if taken by itself, just because it is incomplete and one-sided. Since they are pictures of the same thing, they are compatible. Their seeming incompatibility is created by our looking at them with an inappropriate preconception." Quite; with the spirit of these comments one can agree whole-heartedly. We need, not a solo recital of just one instrument, but a trio. Let us add a constructivistic cello-part to the Platonic-piano and the formalist-violin ones.

This is not the occasion to discuss our authors' view in any further detail, nor to argue for or against them. Rather, let us attempt now to sketch an integrated theory — in the form of a trio — in which what is of positive interest in the authors' account will be shown to be in harmony with the essential tenets of both logicism and formalism. And room must be made also for what is of pan-mathematical value in intuitionism. Let us dub the result 'constructive formal metalogicism,' for want of a better phrase.

Let us consider the system of ordinal mereology, OM, or of homology, HL discussed in V above, and the representation of mathematics that can be developed therein.

The first item to observe is that OM — or, *mutatis mutandis*, HL — without the addition of any non-logical constants, is itself merely a kind of mereological algebra, in which 'P' takes the place of the usual algebraic identity. OM is essentially a Boolean algebra of individuals extended to include ordinal individuals, Boolean sums of such, and so on. Two quite different axiomatic approaches to this theory may be taken. In one of them, an existence assumption may be adopted in which all the entities expressible in the notation are admitted. Thus we could assume a very strong existence principle, stating that, given any individuals x and y — with perhaps some exceptions — there is an ordinal individual z consisting of x ordered with y. All Boolean sums, negatives, and so on, are also assumed to exist, so that this existence principle, by iteration, would assure the existence of all manner of sums of ordinal individuals, of ordinal individuals whose ordinands are such sums, and so on and on. Note also that this principle assures the existence of *triadic* ordinal individuals, and so on, and hence all manner of triadic relations. And so on, in such wise that the theory would condone as individuals, what we would otherwise describe as entities within a simple theory of types.

The other axiomatic structure would keep the assumptions of existence down to a kind of minimum, countenancing the existence of only such entities, ordinal and other, as are actually needed for specific mathematical purposes. This procedure was followed to some extent in OM, which may no doubt be improved upon to meet this demand more fully. Some such demand is appealing to those who think there is a "too-much-ness" about the set-theoretic or type-theoretic approaches to mathematics — too many entities that are never needed in mathematical praxis, too much assumed axiomatically.

The next step in the development of OM is to extend it so as to include logical semiotics — syntax, semantics, and pragmatics — somewhat perhaps in the manner previously described. From there, we move on to event-theory and from thence to a kind of more or less universal grammar. At each step new primitives and postulates are required to handle the new domain of theory. Especially important here, for the discussion of mathematical praxis, are the gerundive predicates for human acts, especially those for the kinds of acts the mathematician actually performs in his day-to-day work, even in the work of constructing formal systems. First, however, let us observe that the new postulates needed at any stage of development of OM are in principle of the same kind as are needed at any earlier stage. All of them may be stated in terms of the truth-functions and first-order quantifiers. Nothing more is needed by way of a logical underpinning. A good methodological principle here might be: Let us make the best of what we've got. This principle would accord well with the advice of Simmias in the *Phaedo*, 85. "I feel myself ... how hard or rather impossible is the attainment of any certainty about questions such as these, ... For [one] should persevere until he has achieved one of two things: either he should discover, or be taught the truth about them; or, if this be impossible, I would have him take the best and most irrefragable of human theories, and let this be the raft upon which he sails through life — not without risk, as I admit, if he cannot find some divine word which will surely and safely carry him."

The view taken here is that first-order quantification-theory, or its surrogates in relational algebra or combinatoric logic, are the "best and most irrefragable" of logical theories we yet have. If better ones are to come, splendid. Meanwhile, we must be wary of the many false pretenders that clamor to be proclaimed better. All of these, without exception to date, have turned out to be mere applications of the standard quantification-theory in which new ideas — usually of a semiotical kind — are involved, these ideas in turn needing symbolization and axiomatization.[5]

What kinds of acts does a mathematician perform in his daily professional chores? As suggested above, a full classification of these is needed and as much systematic interrelating of them as we can think up. As good a preliminary list as any, and probably better than most, is that of Polya.[6] "First: you have to *understand* the problem. Second: find the connection between the data and the unknown.... You should obtain eventually a *plan* of the solution. Third: *carry out* your plan. Forth: *examine* the solution obtained." Each of these in turn can be further broken down: "... look around for an appropriate related problem, Work backwards, Work forwards, Narrow the condition, Widen the condition, Seek a counterexample, Guess and test, Divide and conquer, Change the conceptual mode." And so on and on.

The important point to note here is that all of these activities can be expressed gerundively. To consider only the first, let

(1) '<Und>*e*'

express that *e* is an act of *understanding*, in essentially the sense no doubt of Frege's *apprehension*. We can then add a clause expressing the *object* of the act *e* of understanding.

(2) '(<Und>*e* . *e* $\mathrm{Of}_{\mathrm{Object}}$ *a*)'

thus expresses that *e* is an act of understanding the statement *a* of the given mathematical language, *a* being the problematic statement, the statement concerning which a problem has arisen in the first place. Having an appropriate symbolization for acts of understanding, we can now move on to try to develop a logic of it. Of course this is very difficult to do adequately, and should take account of cognition-theory, a theory of concepts, and the like, about which, alas, we know much less than we would like.

And similarly for each of the other types of mathematical acts and their interrelations.

The reader may object that the symbolizations given by (1) and (2) accomplish little in the way of furtherance of theory. Of course it is the theory that needs developing. In set-theory the notation

(3) '$x \, \varepsilon \, y$',

purporting to express that *x* is a member of *y*, is of little interest aside from a theory characterizing it in some appropriate way. It is only *after* a suitable symbolization is achieved that it is fruitful to try to develop a systematic theory. Axiomatic set-theory would never have gotten very far without 'is an element of', 'is a member of', or 'ε', or the like, taken as a basic notion. In similar vein, not much theory of mathematical praxis can be developed without a slew of symbolizations of the kind (1) and (2) or alternatives thereof. Most of this development remains of course for the future. The beginnings, however, are contained in some of the notions of event-logic.

Just as the basic mereology of OM was extended to gain mathematics, and then further extended to gain syntax, semantics, and so on, so it may be further extended to gain given domains of mathematical physics. The technical problems here are often formidable, as indeed they are at every stage of development. The technical problems faced by Zermelo and Russell, by Hilbert and his followers, and by Brouwer and his, were also formidable. Nothing less than formidable difficulty should be expected when we turn to simplifications of or alternatives to their work or to an attempt to unify it within a single framework such as an extended OM.

The greatest single feature of Hilbert's program was no doubt that it called attention for the first time to mathematical problems of a metatheoretic nature — about a theory or system taken as a whole, or about features or items of a theory or system as a whole in their interrelation with other features or items. What is of permanent value in this approach is now handled in terms of syntax and semantics, both of which can in turn be given systematic formulations. Without such formulations we never really know what the subject is supposed to amount to — what its objects are or what its assumptions are. There is thus no longer any need whatsoever for any quarrel between "logicism" and "formalism." The desirable features of both can now be incorporated within a unified framework such as the extended EM, or within an extended type- or set-theory.

What now about Brouwerian intuitionism, still a highly controversial subject? Let us reflect upon this very briefly from the point of view of the event-theoretical part of OM. This will be in terms of specific kinds of mathematical *acts*, especially acts of *construction*. Constructibility, the notion of an act of constructing a number or function or proof, is no doubt the key notion of intuitionism. The problem for the intuitionist is for him to tell us precisely what types of constructions he finds acceptable or not. Thus we may let

(4) '<Constr>*e*'

express that *e* is an act of constructing something in accord with the criteria of acceptability adopted.[7] The whole theory may then be viewed as built up in terms of expressions of the form (4), plus perhaps others to be supplied. The connection of intuitionism with the theory of computer-programming is also of interest and can presumably be handled from the point of view here.[8]

If these various reflections are sound, we have then in the extended

OM a systematic framework in which the trio of logicism, formalism, and constructivism may play together harmoniously, each contributing to the ensemble the needed ingredient not supplied by any other.

CHAPTER THIRTEEN

On relational domains, the algebra of relations, and relational-term logic

"Mathematics is tricks." Henry Pollock
"No, mathematics is thought."
Abner Shimony

Well, perhaps the appropriate addendum to the foregoing dialogue is to remark that mathematics is thought about tricks and the reasons for them — about the thousands upon thousands of minutiae that make up mathematical language when stripped to its ultimate notions, in the manner, say of *Principia Mathematica* and allied systems. Of particular importance here are the technicalia of quantification-theory, as formidable a bag of tricks as has ever been thought up. That this is the case becomes especially evident when that theory is looked at in the light of other theories that purport to take its place in some fashion or other — such as relational algebra, combinatory logic, or predicate-functor logic.

In a previous paper[1] a preliminary sketch of a pure algebra of n-adic relations — call it 'RA' — was put forward, 'pure' in the sense that no set-theory was made use of in any way at either the object- or metalinguistic level. This kind of formulation is in marked contrast with previous formulations. A purported interpretation of the theory was given in the theory of virtual classes and relations as based on the ordinary first-order theory of quantification.[2] No axiomatization of RA was given, however, and only a feeble attempt was made to show how quantification-theory itself is contained within it in a kind of notational disguise. Let us attempt now, in the present paper, to formulate a simpler kind of theory as a surrogate for it.

Let us recall, first, some basic features of RA itself and of the theory of virtual relations on which it is based. A supply of primitive non-logical relational constants, each of specified finite degree, is presupposed. The algebra RA is thus an *applied* one in this sense (but pure of course in presupposing no set-theory). Boolean operations of forming *sums, products*, and *negations* of n-adic relations are also presupposed, as well as the *universal* and *null* relations of each degree n, where $n \geqq 1$. Where $n = 1$, we speak of a *monadic* relation, as is more or less customary. In addition, an operation for the *Cartesian product* of two relations is needed, as well as a notation for all the *domains* of a relation. The notion of domain here is the generalized one appropriate for n-adic relations. These latter two notions seem not to have been studied very much for their own sake and in generalized form, but are of the highest interest and importance philosophically.

These various notions may be characterized in terms of virtual relations as follows. Let

$$\{x_1 \ldots x_n \ni \text{--}x_1\text{--}\ldots\text{--}x_n\text{--}\}$$

be the virtual relation among $x_1, \ldots, x_n$ where '$\text{--}x_1\text{--}\ldots\text{--}x_n\text{--}$', some formula of quantification-theory constructed in terms of the non-logical relational primitives, holds. We can then let

$$\ulcorner(R_n \cup S_n)_n\urcorner \text{ abbreviate } \ulcorner\{x_1 \ldots x_n \ni (R_n\, x_1 \ldots x_n \vee S_n\, x_1 \ldots x_n)\}\urcorner,$$
$$\ulcorner(R_n \cap S_n)_n\urcorner \text{ abbreviate } \ulcorner\{x_1 \ldots x_n \ni (R_n\, x_1 \ldots x_n \,.\, S_n\, x_1 \ldots x_n)\}\urcorner,$$
$$\ulcorner\text{-}R_n\urcorner \text{ abbreviate } \ulcorner\{x_1 \ldots x_n \ni \sim R_n\, x_1 \ldots x_n\}\urcorner,$$
$$\text{`}V_n\text{'} \text{ abbreviate } \text{`}\{x_1 \ldots x_n \ni (x_1 = x_1 \,.\, x_2 = x_2 \,.\, \ldots \,.\, x_n = x_n)\}\text{'}$$

and

$$\text{`}\Lambda_n\text{'} \text{ abbreviate } \text{`-}V_n\text{'}.$$

These abbreviations, within the theory of virtual classes and relations, introduce the respective Boolean notions. The Cartesian product may be defined by letting

$$\ulcorner(R_n \times S_m)_{(n+m)}\urcorner \text{ abbreviate } \ulcorner\{x_1 \ldots x_n\ y_1 \ldots y_m \ni (R_n x_1 \ldots x_n \,.\, S_m y_1 \ldots y_m)\}\urcorner.$$

Note that the Boolean operations are *homeoadic* in the sense that the result of the operation is of the same degree n as that of the operands, and that the two operands (of the sums and products) are of the same degree n. The degree of the Cartesian product, however, is *cumulative* in the sense that the degree $(n+m)$ of the result of the operation is the (arithmetical) sum of the degrees n and m of the two operands. The notion of the identity of individuals used here may be taken either as a primitive (as in quantification-theory with identity) or else defined, say, in the manner of Hilbert and Bernays.[3] Let it be represented here by 'Id_2', no matter how introduced.

The explanations in the foregoing paragraph are of course merely heuristic, and play no role in RA itself. They merely help us to see how its primitives are to be interpreted. And similarly now for the generalized notion of the domain of a relation.

For a dyadic relation $\boldsymbol{R}_2$, the first domain is merely

(1) $\{x \ni (\mathrm{E}y)\boldsymbol{R}_2\,xy\}$,

and the second or converse domain is

(2) $\{y \ni (\mathrm{E}x)\boldsymbol{R}_2\,xy\}$.

For a triadic relation $\boldsymbol{R}_3$, we must provide not only for

(3) $\{x \ni (\mathrm{E}y)(\mathrm{E}z)\boldsymbol{R}_3\,xyz\}$,

(4) $\{y \ni (\mathrm{E}x)(\mathrm{E}z)\boldsymbol{R}_3\,xyz\}$,

and

(5) $\{z \ni (\mathrm{E}x)(\mathrm{E}y)\boldsymbol{R}_3\,xyz\}$,

but also for

(6) $\{xy \ni (\mathrm{E}z)\boldsymbol{R}_3\,xyz\}$,

(7) $\{xz \ni (\mathrm{E}y)\boldsymbol{R}_3\,xyz\}$,

and

(8) $\{yz \ni (\mathrm{E}x)\boldsymbol{R}_3\,xyz\}$.

The first of these we may think of as the domain, the third as the converse domain, and the second as the middle domain. The other three are *relational* domains, but no obvious terminology for them is available. In the case of tetradic (or quadratic) relations, the situation is still more complicated. To provide for all these, and in general for all the domains including the relational ones, let us let

> $\ulcorner(\mathrm{D}_{i,\ldots,h}\boldsymbol{R}_n)_j\urcorner$ abbreviate $\ulcorner\{x_i\ldots x_h \ni (\mathrm{E}x_k)\ldots(\mathrm{E}x_m)\boldsymbol{R}_n x_1\ldots x_n\}\urcorner$, where $i, \ldots, h, k, \ldots, m$ are any n distinct positive integers (each $\leqq n$), $i, \ldots, h$ are just j in number ($j \geqq 1$), and $i, \ldots, h$ and $k, \ldots, m$ are taken in any order of magnitude, for $n \geqq 2$.

We note now how (1)-(8) may be represented or symbolized in terms of 'D'. (1) in fact is merely $(\mathrm{D}_1\boldsymbol{R}_2)_1$ and (2) is $(\mathrm{D}_2\boldsymbol{R}_2)_1$. (3) is $(\mathrm{D}_1\boldsymbol{R}_3)_1$, (4) is $(\mathrm{D}_2\boldsymbol{R}_3)_1$, (5) is $(\mathrm{D}_3\boldsymbol{R}_3)_1$. (6), (7), and (8) are themselves dyadic relations represented by '$(\mathrm{D}_{1,2}\boldsymbol{R}_3)_2$', '$(\mathrm{D}_{1,3}\boldsymbol{R}_3)_2$' and $(\mathrm{D}_{2,3}\boldsymbol{R}_3)_2$' respectively. And so on for relations of higher degree and their domains, relational and other.

Let us note that expressions such as ‘$(\mathrm{D}_{1,2,3 R3})_3$’ are also significant, so that a relation is allowed to be one of its own domains. But so is ‘$(\mathrm{D}_{1,3,2R3})_3$’ significant, and ‘$(\mathrm{D}_{2,1,3}\boldsymbol{R}_3)_3$’, ‘$(\mathrm{D}_{2,3,1}\boldsymbol{R}_3)_3$’, ‘$(\mathrm{D}_{3,1,2}\boldsymbol{R}_3)$’, and ‘$(\mathrm{D}_{3,2,1}\boldsymbol{R}_3)_3$’. Each of these represents one of the *converses* of the triadic relation $\boldsymbol{R}_3$, just as ‘$(\mathrm{D}_{2,1}\boldsymbol{R}_2)_2$’ represents the converse of $\boldsymbol{R}_2$. And similarly for relations of higher degree, so that the general theory of converses included in the general theory of domains.

Note, by way of a few further examples, that

(9) $\ulcorner\{x \ni (y)\boldsymbol{R}_2\, xy\}\urcorner$

is represented here by

$$\ulcorner\text{-}(\mathrm{D}_1\text{-}\boldsymbol{R}_2)\urcorner,$$

(10) $\ulcorner\{x \ni (\mathrm{E}y)(z)(w)(\mathrm{E}u)(v)\boldsymbol{R}_6\, xyzwuv\}\urcorner$

by

$$\ulcorner(\mathrm{D}_1\text{-}(\mathrm{D}_{1,2}\text{-}(\mathrm{D}_{1,2,3,4}\text{-}(\mathrm{D}_{1,2,3,4,5}\text{-}\boldsymbol{R}_6)_5)_4)_2)_1\urcorner,$$

(11) $\ulcorner\{xwyz \ni \boldsymbol{R}_4\, xyzw\}\urcorner$

by

$$\ulcorner(\mathrm{D}_{1,4,2,3}\boldsymbol{R}_4)_4\urcorner,$$

(12) $\ulcorner\{xy \ni (\mathrm{E}z)(\mathrm{E}w)(u)(\mathrm{E}v)\boldsymbol{R}_6\, xvyuwz\}\urcorner$

by

$$\ulcorner(\mathrm{D}_{1,2}\text{-}(\mathrm{D}_{1,2,3,4}\text{-}(\mathrm{D}_{1,2,3,4,5}(\mathrm{D}_{1,6,2,5,4,3}\boldsymbol{R}_6)_6)_5)_4)_2\urcorner.$$

In the formulation of RA, all talk of virtual classes and relations is, of course, dropped, the effect of such being achieved by use of the Boolean notions, Cartesian products, and domains. In an algebra there is always a primitive ‘=’ for identity, so that $\ulcorner\boldsymbol{R}_n = \boldsymbol{S}_n\urcorner$ is the only kind of atomic formula admitted, but with ‘∼’ and ‘v’, say, available as truth-functional primitives to provide for molecular formulae. The basic principles of RA — and, ultimately the axioms — must provide for the truth-functions, for identity as between *n*-adic relations, for the Boolean notions, and for Cartesian products and domains. A list of some of the principles needed was put forward in the previous paper. Not all of these are required as axioms, although some additional ones are.

In the original presentation of RA in *Mind, Modality, Meaning, and Method*, the sentence

(13) '$(\mathrm{E}x)(y)(\mathrm{E}u)(v)\boldsymbol{T}_4\, xyuv$'

was said to be expressible by

(14) '$\sim$ -$(\mathrm{D}_{1,3}\text{-}\boldsymbol{T}_4)_2 = \Lambda_2$'.

But unfortunately this is not correct. This latter states, in virtual-relation terms, that

$$\sim \{xu \ni (\mathrm{E}y)(\mathrm{E}v)\text{-}T_4 xyuv\} = \Lambda_1,$$

or, equivalently, that

(15) $(\mathrm{E}x)(\mathrm{E}u)(y)(v)\text{-}\boldsymbol{T}_4\, xyuv$.

But of course (15) is not equivalent to (13), although it logically implies it. The form for (13) should be rather

$$\text{`-}(\mathrm{D}_1\text{-}(\mathrm{D}_{1,2}\text{-}(\mathrm{D}_{1,2,3}\text{-}\boldsymbol{T}_4)_3)_2)_1 = \Lambda_1\text{'},$$

as may easily be verified. And, similarly, the Axiom of Pairs of set-theory has the form

$$\text{`}(\mathrm{D}_{1,2}\text{-}(\mathrm{D}_{1,2,3}\text{-}(((\mathrm{D}_{1,4,2,3}(\mathrm{E}_2 \times \mathrm{V}_2)_4)_4 \cap ((\mathrm{Id}_2 \times \mathrm{V}_2)_4 \cup (\mathrm{D}_{1,3,2,4}(\mathrm{Id}_2 \times \mathrm{V}_2)_4)_4)_4)_4 \cup (\text{-}(\mathrm{D}_{1,4,2,3}(\mathrm{E}_2 \times \mathrm{V}_2)_4)_4 \cap (\text{-}(\mathrm{Id}_2 \times \mathrm{V}_2)_4 \cap \text{-}(\mathrm{D}_{1,3,2,4}(\mathrm{Id}_2 \times \mathrm{V}_2)_4)_4)_4)_4)_3)_2 = \mathrm{V}_2\text{'},$$

and the given instances of the *Aussonderungsschema* the form

$$\ulcorner(\mathrm{D}_1\text{-}(\mathrm{D}_{1,2}\text{-}(((\mathrm{E}_2 \times \mathrm{V}_1)_3 \cap ((\boldsymbol{R}_1 \times \mathrm{V}_2)_3 \cap (\mathrm{D}_{1,3,2}(\mathrm{E}_2 \times \mathrm{V}_1)_3)_3)_3)_3 \cup (\text{-}(\mathrm{E}_2 \times \mathrm{V}_1)_3 \cap (\text{-}(\boldsymbol{R}_1 \times \mathrm{V}_2)_3 \cup \text{-}(\mathrm{D}_{1,3,2}(\mathrm{E}_2 \times \mathrm{V}_1)_3)_3)_3)_3)_3)_2)_1 = \mathrm{V}_1\urcorner.$$

(Please note also that, on page 96 of *Mind, Modality, Meaning, and Method,* in the second lines of the definitions of '/' and '#' the $\ulcorner(\cup S_2)_2\urcorner$ should be replaced by S_2. And on p. 95, $\ulcorner\sim T_m = \Lambda_m\urcorner$ should be added as a conjunct to the hypotheses of *Pr28* and *Pr29*.)

Given an *n*-adic relation, for $n \geqq 2$, the totality of its converses together with all its domains constitutes a family of closely *affiliated* relations. All of the affiliates are constructed from the given one and are recognized as relations along with it. In some sense, they are "given" along with the original one, merely awaiting a proper notation, as it were. Once we have such a notation, RA is seen to be merely an extension of an essentially Boolean theory as augmented with Cartesian products. The admission of the relations affiliated with a given relation thus seems not only natural, but a small price to pay for the algebraic richness forthcoming as a result. Further, RA itself may now be seen to be wholly Booleanized, so to speak, all erstwhile relational notions — instrinsically relational or non-Boolean ones — being now assimilated in the theory of domains.

It should be observed that '=' for identity as between n-adic relations in RA is a very strong primitive indeed, and the two axiom-schemata governing it are very strong principles. These are that

$$\vdash \boldsymbol{R}_n = \boldsymbol{R}_n,$$

and

$\vdash \boldsymbol{R}_n = \boldsymbol{S}_n \supset {-}{-}\boldsymbol{R}_n{-}{-} = {-}{-}\boldsymbol{S}_n{-}{-}$, where $\ulcorner{-}{-}\boldsymbol{S}_n{-}{-}\urcorner$ differs from the term $\ulcorner{-}{-}\boldsymbol{R}_n{-}{-}\urcorner$ only in containing occurrences of $\boldsymbol{S}_n$ in one or more places where there are occurrences of $\boldsymbol{R}_n$ in $\ulcorner{-}{-}\boldsymbol{R}_n{-}{-}\urcorner$.

It is interesting to observe that the effect of having identity may be achieved in a much more economical way. The result will no longer be an algebra, however, but a special kind of logic of relations.

We should observe also that the truth-functional connectives may be eliminated as primitives of RA without loss. Recall the principles that

$$\vdash \sim \boldsymbol{R}_n = \mathrm{V}_n \equiv (\mathrm{D}_1(\mathrm{V}_1 \times -\boldsymbol{R}_n)_{(n+1)})_1 = \mathrm{V}_1,$$

$$\vdash (\boldsymbol{R}_n = \mathrm{V}_n \,.\, \boldsymbol{S}_n = V_n) \equiv (\boldsymbol{R}_n \cap \boldsymbol{S}_n)_n = \mathrm{V}_n,$$

and

$$\vdash (\sim \boldsymbol{R}_n = \mathrm{V}_n \vee \sim \boldsymbol{S}_n = \mathrm{V}_n) \equiv (\mathrm{D}_1(\mathrm{V}_1 \times -(\boldsymbol{R}_n \cap \boldsymbol{S}_n)_n)_{(n+1)})_1 = \mathrm{V}_1,$$

or, better,

$$\vdash (\boldsymbol{R}_n = \mathrm{V}_n \vee \boldsymbol{S}_n = \mathrm{V}_n) \equiv (-(\mathrm{D}_1(\mathrm{V}_1 \times -\boldsymbol{R}_n)_{(n+1)})_1 \cup -(\mathrm{D}_1(\mathrm{V}_1 \times -\boldsymbol{S}_n)_{(n+1)})_1)_1 = \mathrm{V}_1.$$

Whenever, in RA, we wish to use a truth-functional connective, we may elminate it in view of these equivalences.

It should be observed also that the only atomic formulae we need consider are of the form $\ulcorner T_n = \mathrm{V}_n\urcorner$ in view of the Boolean principle that

$$\vdash \boldsymbol{R}_n = \boldsymbol{S}_n \equiv ((-\boldsymbol{R}_n \cup \boldsymbol{S}_n)_n \cap (\boldsymbol{R}_n \cup -\boldsymbol{S}_n)_n)_n = \mathrm{V}_n.$$

It is true that '=' occurs in this equivalence on the right-hand side. In place of $\ulcorner T_n = \mathrm{V}_n\urcorner$, let us now write merely

$$\ulcorner \mathrm{V}_n T_n \urcorner,$$

to the effect that such and such an n-adic relation is *universal* in the sense appropriate to n-adic relations. No properties of identity as such need then be postulated, as we shall see in a moment. Actually, we can simplify RA in another respect also, by using primitively only Cartesian products of the form ⌜$(\boldsymbol{R}_n \times \mathrm{V}_1)_{(n+1)}$⌝, where V_1 is the universal monadic relation. We shall see how the full effect of having Cartesian products may be achieved in this way.

Let us go on, now, to formulate the essentials of RA, but without using identity or the truth-functions. As primitives, let us take '-' for Boolean negation, '∪' for Boolean sums, '×' for the restricted Cartesian products, and 'D' with suitable subscripts for relational domains. Erstwhile individual constants are to be handled predicatively, so that 'Soc', for example, may represent the predicate 'Socratizes'. At least one such primitive primitive is presupposed, say 'Soc' itself.

The following recursive specification is of the general notion of being a *relational term of degree n* where $n \geqq 1$.

1. 'Soc' (and any other primitive for an erstwhile proper name) is a relational term of degree 1.
2. If $\boldsymbol{R}_n$ is a relational term of degree n, so is ⌜$-\boldsymbol{R}_n$⌝.
3. If $\boldsymbol{R}_n$ and $\boldsymbol{S}_n$ are relational terms of degree n, so is ⌜$(\boldsymbol{R}_n \cup \boldsymbol{S}_n)_n$⌝.
4. If $\boldsymbol{R}_n$ is a relational term of degree n, ⌜$(\boldsymbol{R}_n \times (\mathrm{Soc} \cup -\mathrm{Soc})_1)_{(n+1)}$⌝ is a relational term of degree $(n+1)$.
5. If $\boldsymbol{R}_{(n+m)}$ is a relational term of degree $(n+m)$, ⌜$(\mathrm{D}_{i,\ldots,j}\, \boldsymbol{R}_{(n+m)})_n$⌝ is a relational term of degree n, where (as needed).

To facilitate the notation, we may let

D1. 'V_1' abbreviate '$(\mathrm{Soc} \cup -\mathrm{Soc})_1$',

and then

D2a. 'V_2' abbreviate '$(\mathrm{V}_1 \times \mathrm{V}_1)_2$',

D2b. 'V_3' abbreviate '$(\mathrm{V}_2 \times \mathrm{V}_1)_2$',

and so on. In addition we may let

D3a. ⌜$(\boldsymbol{R}_n \times \mathrm{V}_2)_{(n+2)}$⌝ abbreviate ⌜$((\boldsymbol{R}_\mathrm{n} \times \mathrm{V}_1)_{(n+1)} \times \mathrm{V}_1)_{(n+2)}$⌝,

D3b. ⌜$(\boldsymbol{R}_n \times \mathrm{V}_3)_{(n+3)}$⌝ abbreviate ⌜$((\boldsymbol{R}_n \times \mathrm{V}_2)_{(n+2)} \times \mathrm{V}_1)_{(n+3)}$⌝,

and so on. Also

D4. ⌜$(\mathrm{V}_n \times \boldsymbol{R}_m)_{(n+m)}$⌝ abbreviates ⌜$(\mathrm{D}_{(m+1),\ldots,(m+n),1,\ldots,m}(\boldsymbol{R}_m \times \mathrm{V}_n)_{(n+m)})_{(n+m)}$⌝,

D5. ⌜$(\boldsymbol{R}_n \times \boldsymbol{S}_m)_{(n+m)}$⌝ abbreviates ⌜$((\boldsymbol{R}_n \times \mathrm{V}_m)_{(n+m)} \cap (\mathrm{V}_n \times \boldsymbol{S}_m)_{(n+m)})_{(n+m)}$⌝.

In this way we can achieve the full effect of having Cartesian products.

By a *formula* let us now understand any expression of the form $\ulcorner \mathrm{V}_n \, \boldsymbol{R}_n \urcorner$ where $\boldsymbol{R}_n$ is a relational term of degree *n*. The formulae are to be understood as saying that such and such a relation of degree *n* has the kind of universality appropriate to *n*-addic relations. The only expressions allowed in this kind of logic are the relational terms. Let us therefore call it 'relational-term logic', or for short, 'RTL'.

A few further abbreviations are useful. We may let

D6. $\ulcorner(\boldsymbol{R}_n \cap \boldsymbol{S}_n)_n\urcorner$ be short for $\ulcorner\text{-}(\text{-}\boldsymbol{R}_n \cup \text{-}\boldsymbol{S}_n)_n\urcorner$,

D7a. 'Λ_1' for '$\text{-}\mathrm{V}_1$',

D7b. 'Λ_2' for '$(\Lambda_1 \times \Lambda_1)_2$',

and so on,

D8. $\ulcorner(\boldsymbol{R}_n \supset \boldsymbol{S}_n)_n\urcorner$ for $\ulcorner(\text{-}\boldsymbol{R}_n \cup \boldsymbol{S}_n)_n\urcorner$,

and

D9. $\ulcorner(\boldsymbol{R}_n \Box \boldsymbol{S}_n)_n\urcorner$ for $\ulcorner((\boldsymbol{R}_n \supset \boldsymbol{S}_n)_n \cap (\boldsymbol{S}_n \supset \boldsymbol{R}_n)_n)_n\urcorner$.

D6 and *D7* are definitions of a familiar kind, *D8* and *D9* less so. *D8* introduces the notion of the Boolean *implexion* of *n*-adic relations, and *D9* that of the Boolean *identification* of such. In terms of these the notions of *inclusion* and *identity* of *n*-adic relations may be introduced, by letting

D10. $\ulcorner\boldsymbol{R}_n \subset \boldsymbol{S}_n\urcorner$ abbreviate $\ulcorner\mathrm{V}_n \, (\boldsymbol{R}_n \supset \boldsymbol{S}_n)_n\urcorner$.

and

D11. $\ulcorner\boldsymbol{R}_n = \boldsymbol{S}_n\urcorner$ abbreviate $\ulcorner\mathrm{V}_n \, (\boldsymbol{R}_n \Box \boldsymbol{S}_n)_n\urcorner$.

Let us go on now to list some fundamental principles for RTL as follows. We first have some Boolean ones.

Pr1. $\vdash (\boldsymbol{R}_n \cup \boldsymbol{R}_n)_n \subset \boldsymbol{R}_n$,

Pr2. $\vdash \boldsymbol{R}_n \subset (\boldsymbol{R}_n \cup \boldsymbol{S}_n)_n$,

Pr3. $\vdash (\boldsymbol{R}_n \cup \boldsymbol{S}_n)_n \subset (\boldsymbol{S}_n \cup \boldsymbol{R}_n)_n$,

Pr4. $\vdash (\boldsymbol{R}_n \supset \boldsymbol{S}_n)_n \subset ((\boldsymbol{T}_n \cup \boldsymbol{R}_n)_n \supset (\boldsymbol{T}_n \cup \boldsymbol{S}_n)_n)_n$,

MP. If $\vdash \mathrm{V}_n \, \boldsymbol{R}_n \subset \boldsymbol{S}_n$, then $\vdash \mathrm{V}_n \, \boldsymbol{S}_n$.

These principles will be recognized as sufficient for providing for a Boolean algebra of *n*-adic relations — together with an existence assumption that will be given in a moment. *MP* is of course the adaptation of *modus ponens* needed for *n*-adic relations.

The following principles govern the restricted Cartesian products and the relational domains.

Pr5. $\vdash (-\boldsymbol{R}_n \times \mathrm{V}_1)_{(n+1)} = -(\boldsymbol{R}_n \times \mathrm{V}_1)_{(n+1)}$,

Pr6. $\vdash ((\boldsymbol{R}_n \cup \boldsymbol{S}_n)_n \times \mathrm{V}_1)_{(n+1)} = ((\boldsymbol{R}_n \times \mathrm{V}_1)_{(n+1)} \cup (\boldsymbol{S}_n \times \mathrm{V}_1)_{(n+1)})_{(n+1)}$,

Pr7a. $\vdash (\mathrm{D}_{1,\ldots,n}(\boldsymbol{R}_n \times \mathrm{V}_1)_{(n+1)})_n = \boldsymbol{R}_n$,

Pr7b. $\vdash -(\mathrm{D}_{1,\ldots,n}(-\boldsymbol{R}_n \times \mathrm{V}_1)_{(n+1)})_n = \boldsymbol{R}_n$,

Pr8. $\vdash -(\mathrm{D}_{1,\ldots,n}-(\boldsymbol{R}_{(n+1)} \supset \boldsymbol{S}_{(n+1)})_{n+1)})_n \subset (-(\mathrm{D}_{1,\ldots,n}-\boldsymbol{R}_{(n+1)})_n \supset -(\mathrm{D}_{1,\ldots n}$ $-\boldsymbol{S}_{(n+1)})_n)_n$,

Pr9. $\vdash -(\mathrm{D}_{1,\ldots,(n-1),(n+1)}-(\boldsymbol{R}_n \times \mathrm{V}_1)_{(n+1)})_n \subset -(\mathrm{D}_{1,\ldots,n}(\mathrm{D}_{1,\ldots,(n-1),(n+1),n}-(\boldsymbol{R}_n$ $\times \mathrm{V}_1)_{(n+1)})_{(n+1)})_n$,

Pr10. $\vdash -(\mathrm{D}_{1,\ldots,(n-1),(n+1)}-(\boldsymbol{R}_n \times \alpha_1)_{(n+1)})_n \subset -(\mathrm{D}_{1,\ldots n}(\mathrm{D}_{1,\ldots,(n-1),(n+1),n}-(\boldsymbol{R}_n$ $\times \mathrm{V}_1)_{(n+1)})_{(n+1)})_n$, where α_1 is any primitive for an individual constant,

Gen. If $\vdash \mathrm{V}_n\, \boldsymbol{R}_n$ then $\vdash \mathrm{V}_1\ -(\mathrm{D}_{(n+1)}(-\boldsymbol{R}_n \times \mathrm{V}_1)_{(n+1)})_1$.

We have also some principles governing converses, as follows.

Pr11. $\vdash (\mathrm{D}_{i,\ldots,j}\,\boldsymbol{R}_n = -(\mathrm{D}_{i,\ldots,j}-\boldsymbol{R}_n)_n$, where $i, \ldots, j$ are any n distinct positive integers, each $\leqq n$,

Pr12. $\vdash (\mathrm{D}_{i,\ldots,j}(\boldsymbol{R}_n \cup \boldsymbol{S}_n)_n)_n \subset ((\mathrm{D}_{i,\ldots,j}\,\boldsymbol{R}_n)_n \cup (\mathrm{D}_{i,\ldots,j}\,\boldsymbol{S}_n)_n)_n$, where (etc.),

Pr13. $\vdash (\boldsymbol{R}_n \supset \boldsymbol{S}_n)_n \subset ((\mathrm{D}_{i,\ldots,j}\,\boldsymbol{R}_n)_n \supset (\mathrm{D}_{i,\ldots,j}\,\boldsymbol{S}_n)_n)_n$,

Pr14a. $\vdash (\mathrm{D}_{2,1}(\mathrm{D}_{2,1}\boldsymbol{R}_2)_2)_2 = \boldsymbol{R}_2$,

Pr14b. $\vdash (\mathrm{D}_{1,2}(\mathrm{D}_{1,2}\boldsymbol{R}_2)_2)_2 = \boldsymbol{R}_2$,

Pr14c. $\vdash (\mathrm{D}_{1,3,2}(\mathrm{D}_{1,3,2}\boldsymbol{R}_3)_3)_3 = \boldsymbol{R}_3$,

Pr14d. $\vdash (\mathrm{D}_{2,1,3}(\mathrm{D}_{2,1,3}\boldsymbol{R}_3)_3)_3 = \boldsymbol{R}_3$,

Pr14e. $\vdash (\mathrm{D}_{3,1,2}(\mathrm{D}_{2,3,1}\boldsymbol{R}_3)_3)_3 = \boldsymbol{R}_3$,

Pr14f. $\vdash (\mathrm{D}_{2,3,1}(\mathrm{D}_{3,1,2}\boldsymbol{R}_3)_3)_3 = \boldsymbol{R}_3$,

Pr14g. $\vdash (\mathrm{D}_{3,2,1}(\mathrm{D}_{3,2,1}\boldsymbol{R}_3)_3)_3 = \boldsymbol{R}_3$,

and so on for relations of higher degree.

Finally, we need a general principle concerning the uniqueness and existence of the monadic relation Soc and other relations representing erstwhile individual constants.

Pr15. $\vdash \alpha = \text{-}(D_1\text{-}((\alpha \times V_1)_2 \cap ((D_{2,1}(\alpha \times V_1)_2 \supset Id_2)_2)_2)_1$, where α is any primitive for an individual constant (and Id_2 is the relation of identity of individuals).

These principles are by no means exhaustive but are given as mere samples of principles that would be needed in any full axiomatization.

We need not tarry here with the deductive development of RTL. Much work yet remains to be done to achieve the minimum axioms sufficing for the theory — to say nothing of stating precisely what it means to say this and then proving that it obtains. To do this last, in a satisfactory way, would require a semantical metalanguage, with its attendant axioms and rules, in which the proof could be carried out. A very considerable effort would be required to formulate such a metalanguage for such a proof, and many unforseen technical difficulties would no doubt arise. The same of course is to be said for all so-called "proofs" of consistency, completeness and independence.[5]

An objection that might be raised against the notation for domains used here is that it makes abundant use of numerical subscripts and therefore the theories here presuppose arithmetic. But no. The numerical subscripts are used as mere notational markers, and their use here is as mere syntactical abbreviation, as would become pellucid if the syntax were spelled out in detail in terms of concatenation and shape-descriptions.[6] Also it might be objected that, although variables such as '*x*', '*y*', and so on, are not needed in RA or RTL, abundant use is made of '$\boldsymbol{R}_n$', and so on, as syntactical variables. True, but if the syntax, again, were fully spelled out, it in turn could be viewed as an applied relational-term logic without syntactical variables.

Let us note now relational domains are of use in giving logical forms to certain natural sentences. The first person ever to make use of them, or even to be aware of their existence, was probably C.S. Peirce. Let us turn therefore to an important paper of his of 1870.[7] This was long before he became aware of the quantifiers, and it is interesting to note that he felt the need of generalized domains to handle certain natural noun-phrases.

Consider, by way merely of an example, the sentence that "every lover of somebody who is a servant to nothing but a woman stands to nothing but women in the relation of lover of nothing but a servant of them." The two noun-phrases needed here may readily be symbolized in virtual-relation terms, and hence in RA or RTL. Let 'L_2' and 'S_2' stand for the dyadic relation of loving and being a servant of, respectively, and let 'W_1' stand for the class of women. Then the required noun-phrases, in quantificational terms, are

$$`\{x \ni (Ey)((z)(S_2yz \supset W_1z) \,.\, L_2xy)\}'$$

and

$$`\{x \ni (y)((z)(L_2xz \supset S_2zy) \supset W_1y\}'.$$

In the symbolism of RTL, these become

$$`(D_1\text{-}(D_{1,2}\text{-}(((V_1 \times S_2)_3 \supset (V_2 \times W_1)_3)_3 \cap (L_2 \times V_1)_3)_3)_2)_1'$$

and

$$`\text{-}(D_1\text{-}(D_{1,2}(D_{1,3,2}(((L_2 \times V_1)_3 \supset (V_1 \times S_2)_3)_3 \supset (V_2 \times W_1)_3)_3)_3)_2)_1'.$$

Another example from Peirce is the sentence "every lover of somebody who is servant to nothing but a woman stands to nothing but women in the relation of lover of nothing but a servant of them." The two noun-phrases here, in the notation of RTL, are respectively,

$$`\text{-}(D_1\text{-}(D_{1,2}(D_{1,3,2}(((L_2 \times V_1)_3 \supset (V_1 \times S_2)_3)_3 \cap (V_1 \times W_1 \times V_1)_3)_3)_3)_2)_1'$$

and

$$`\text{-}(D_1\text{-}(D_{1,2}((L_2 \times V_1)_3 \supset ((V_1 \times S_2)_3 \cap (V_2 \times W_1)_3)_3)_3)_2)_1)'.$$

Peirce's two sentences are then formed by placing the inclusion-sign '$\subset$' between the two relevant phrases. The results, incidentally, are logically true statements, as Peirce in effect observes.

To the uninitiate, these examples may seem somewhat far-fetched. These and similar noun-phrases, however, abound in natural language. It is therefore essential that a method for handling them be available in any general theory of relations adequate for the normal tasks expected of it. It is interesting to observe also that it was for a notation for relational demains of any complexity that Peirce was struggling, in his early papers on the algebra and logic of relations, before coming to terms with the quantifiers in his papers of 1883 and 1885.[8]

Let us go on now to show how RTL is but a step or so removed from a kind of *combinatory* logic of relations, one in which the combinators or operators apply to relations with relations as the result.[9] We shall need just six such combinators, namely,

S, N, P, C, I, and E.

S will provide for Boolean sums of *n*-adic relations, N for negations, P for the restricted Cartesian products, C for *conversion*, I for *inversion*, and E for handling the construction of domains. Conversion and inversion may be explained — again — in terms of virtual relations.

$$\ulcorner(\mathrm{C}\boldsymbol{R}_n)_n\urcorner$$

is to represent

$$\ulcorner\{x_n x_1 \ldots x_{(n\text{-}1)} \ni \boldsymbol{R}_n\, x_1 \ldots x_n\}\urcorner,$$

and

$$\ulcorner(\mathrm{I}\boldsymbol{R}_n)_n\urcorner$$

is to represent

$$\ulcorner\{x_2 x_1 x_3 \ldots x_n \ni \boldsymbol{R}_n\, x_1 \ldots x_n\}\urcorner.$$

E functions as a combinator for forming relational domains. Thus

$$\ulcorner(\mathrm{E}\boldsymbol{R}_n)_{(n\text{-}1)}\urcorner$$

is to represent

$$\ulcorner\{x_1 \ldots x_{(n\text{-}1)} \ni (\mathrm{E}x_n)\boldsymbol{R}_n\, x_1 \ldots x_n\}\urcorner.$$

The use of 'E' here is to suggest the existential quantifier involved.

To see how these combinators function, let us note that (1) above may now be expressed by

$$\ulcorner(\mathrm{E}\boldsymbol{R}_2)_1\urcorner,$$

and (9) by

$$\ulcorner\text{-}(\mathrm{E}\text{-}\boldsymbol{R}_2)_1\urcorner.$$

(11) above becomes

$$\ulcorner(\mathrm{I}(\mathrm{C}\boldsymbol{R}_n)_n)_n\urcorner,$$

(10) becomes

$$\ulcorner(\mathrm{E}\text{-}(\mathrm{E}(\mathrm{E}\text{-}(\mathrm{E}\text{-}(\mathrm{E}\text{-}\boldsymbol{R}_6)_5)_4)_3)_2)_1\urcorner,$$

and (12) becomes

$$\ulcorner(\mathrm{E}(\mathrm{E}(\mathrm{C}(\mathrm{C}(\mathrm{I}\text{-}(\mathrm{E}\text{-}(\mathrm{C}(\mathrm{C}(\mathrm{C}(\mathrm{E}(\mathrm{C}(\mathrm{C}(\mathrm{C}(\mathrm{C}\boldsymbol{R}_6)_6)_6)_6)_5)_5)_5)_5)_4)_4)_4)_4)_4)_3)_2\urcorner,$$

and so on.

The other combinators are such that

$$\ulcorner(\mathrm{S}(\boldsymbol{R}_n,\boldsymbol{S}_n))_n\urcorner,$$

$\ulcorner NR_n \urcorner$,

and

$\ulcorner (PR_n)_{(n+1)} \urcorner$

represent the Boolean sums, negation, and restricted Cartesian products, respectively.

Here, as in RTL, all our formulae may be taken to be of the form $\ulcorner V_n R_n \urcorner$, where '$V_n$' is defined as there. It is easy to see that the principles *Pr1*-*Pr15* may all be expressed as principles in the new notation.

This quasi-combinatory logic of relations is really merely RTL in a kind of notational disguise. Even so, it is just interesting enough perhaps to be given a label of its own. Let us call it 'CLR', the combinatory logic of relations. The use of the letters 'S', 'N', 'P', and so on, is akin to uses of capital roman letters in Curry's formulations. It is interesting, too, to note that CLR, like RA and RTL, can be used as a surrogate for the quantification-theory. In fact, it provides a kind of combinatory theory of quantification.

paper are sharply to be contrasted with the mereological and homological theories of them in V above. However, the theories there could readily be formulated as either an RA or RTL or a CLR, where 'P' and 'Ord' are taken as the only primitive non-logical relation-symbols. The mechanisms of RA or RTL or CLR would merely provide the underlying logic needed in place of the quantification-theory actually used. And similarly for any other theory formulable in quantificational terms.

CHAPTER FOURTEEN

On Husserlian wholes and parts

"Ornare res ipsa negat, contenta docere."

At the recent International Workshop on Parts and Wholes at the University of Lund in June, 1983, Barry Smith called attention to the seminal work on this topic by Husserl and Leśniewski.[1] "The Husserl tradition ... deals not only with ('vertical') relations between parts and their wholes, but also with the ('lateral') relations amongst the parts of a single whole," especially relations of "necessary dependence." "There is an infinite variety of such lateral dependence relations giving rise to an infinite variety of different types of whole which the Leśniewski approach is simply unable to distinguish. Yet Husserl saw that the theory of lateral, part-whole relations can be embedded smoothly into a vertical mereological theory of the Leśniewski type, producing a radical increase both in logical and descriptive power and in complexity of mathematical structure."

In his contribution to the Workshop, Kevin Mulligan[2] likewise claimed "that it is possible to provide a formal theory of part and whole, a theory adequate for the understanding of linguistic structure, as of all other real structure in the spatio-temporal world, and this *without* appeal to the alien and descriptively inadequate apparatus of set theory." Mulligan also makes obeisance to Husserl and Leśniewski. Of Husserl's theory of parts and wholes, he comments that it "is capable of being formalised, though not by any recourse to the standard formalisms ..." of "set theory or the predicate calculus." These latter are "demonstrably inadequate," we are told — but we are not told why. Of Leśniewski we are told merely that he "worked out an intuitively elegant formalisation of a fragment of Husserl's logical grammar," which comprises "much of what is now regarded as belonging to the domain of semantics."

Both of these comments seem somewhat "woolly," from a historical point of view, and not quite accurate. To "see" that such and such a theory is possible is, no doubt, a necessary precondition for an explicit formulation of it. It is true that Leśniewski's mereology is not concerned with "lateral" or "dependence" relations, but it is far from clear that Husserl saw how — or that — a theory of them could be "embedded" in a mereology of Leśniewskian type. On the contrary, it cannot be, the Leśniewskian theory having no truck with intensions (or intentions) in the way the Husserlian theory requires. Nor can the latter be formalized without using the "predicate calculus" or some surrogate thereof. In fact, no theory can be. Also it is much to simple, and indeed misleading, to claim *sans phrase* that "Leśniewski worked out ... a fragment of Husserl's categorial grammar."

The great defect of Husserl's approach to mereology[3], and that of his followers, it would appear, is that it has been content with a mere "seeing" that a theory of it is possible, without the barest hint as to how an explicit formalization of it could be carried out. Let us attempt here to make some suggestions toward remedying this defect. In so doing, let us bear in mind the mereology and/or homology of V above, and concentrate attention on such additional notions as are needed for discussing "dependence" and allied relations. Fundamental here are the so-called *event-descriptive predicates* used previously, together with the full resources of event-logic.[4] (And of course much must be made of the *relations* determining the "dependence" of parts.) Let us begin by following Husserl's discussion more or less step by step, in *Investigation III*, parting from it as required for rigor of theory and precision of presentation. The aim is thus not to formalize Husserl, but merely to show how some of his salient insights may be accommodated within the event-logic framework. Then, at the end, we will be in a position to see that what Husserl and his followers seek to accomplish in terms of "dependence" relations, may be more clearly and efficaciously provided by explicitly symbolizing many structural and quasi-linguistic relations that are more and more being seen to be essential for the logical study of language.

Not everything that is said to be a "part" of something should be properly so regarded, once a rigorous sense of 'part' has been settled on. No "quality" or "property" is a part of a thing, nor (p. 435) a "Species, Genus..., Relation..., Collection, Unity, Number, Series, Ordinal Number, Magnitude, etc.," to cite some of Husserl's "categories." No, let us make no category-mistakes here, but be clear *ab initio* as to the categories we use. Let us think of a thing or "object" a a *cardinal individual* in the sense of V, i.e., as a unit cardinal entity or any Boolean sum of such. The units are the minimalia admitted in the theory, sums of which are ordinary individuals.

"Objects can be related to one another as Wholes to Parts," Husserl writes (p. 436); "they can also be related to one another as coordinated parts of a whole. These sorts of relations [what sorts? we must ask] have an *a priori* foundation in the Idea of an object. Every object is either actually or possibly a part, i.e., there are actual or possible wholes that include it. Not every object, on the other hand, need perhaps have parts, and we have therefore the ideal division of objects into the simple and the complex."

In the theory here the simples are to be the *cardinal units*, and are to have no parts *other than themselves and the null object* N. The basic part-whole relation P here is taken in such a way as to allow every object to bear P to itself. Husserl takes the basic relation so as to exclude this. But we may readily define

'$x\ \mathrm{P_H}\ y$' as '(CardEnt x . CardEnt y . x P y . $\sim x$ P N . $\sim y$ P x)',

to express that x is an *Husserlian part of* y. Also in admitting a null object, we follow Leśniewski rather than Husserl, who seems not to admit such.

The relation of part-to-whole is P, that of whole-to-part is the *converse* of P, usually symbolized as '$^{\cup}$P'. Clearly it holds that

$$\vdash (x)(y)(x\ \mathrm{P}\ y \equiv y\,^{\cup}\mathrm{P}\ x).$$

And by 'actual or possible wholes' we mean all "objects" as spread out before us, as it were, in the Peirce-Frege tense of timelessness.

Husserl is very explict that "the terms 'complex' and 'simple' are ... defined by the qualification of having parts or not having parts." Thus we can let

'Simp x' abbreviate '($\sim$ (Ey)y $\mathrm{P_H}$ x . CardEnt x)'

and

'Comp x' abbreviate '($\sim$ Simp x . CardEnt x)'.

What now about regarding "color and shape as the associated parts of a colored expanse?" Talk of such "goes against linguistic usage," Husserl notes (p. 437). "For in such wholes the parts have relative dependence as regards one another; we find them so closely united as to be called 'interpenetrating'. It is quite different in the case of wholes which are broken up, or could be broken up, into *pieces*. The parts are here not merely disjoined from each other, but relatively independent, they have the character of mutually-put-together *pieces*." Husserl seems willing to go against linguistic usage in noting that "every nonrelative 'real' (*reale*) predicate ... points to a part of the object which is the predicate's subject: 'red' and 'round', e.g., do so, but not 'existent' or 'something'."

To speak of color and shape as "parts" of a colored expanse, either in a visual field or *in natura*, not only goes against linguistic usage but involves a category-mistake as well. The color and shape of an object seem best handled as properties or attributes — attributive relations, we can say here, in the sense of V — and not as "parts." It would thus seem best *not* to use 'part' *ab initio* in the wide sense that Husserl does.

Husserl remarks that "if we now try to pin down the limitations which mark off ... [the] ordinary [usage of 'part'], from *our* notion of part, we come up against the fundamental distinction ... of *independent* and *non-independent* [or dependent] parts." Independent parts may readily be handled by means of 'P'. What we should do now is to try to show, as sympathetically as possible, that Husserl's dependent parts may also be fully handled on this narrower and more natural — and, be it said, more Leśniewskian — construal of 'part'.

Let us now let, as previously[5],

(1) '<Evid> *e*'

express that *e* is an act or "state of conciousness" of *apodictic self-evidence*. '<Evid>' here is an event-descriptive predicate, and expressions of the kind (1) are essential for accommodating Husserl's vocabulary. (1) could also be read: *e* is an act of not-being-able-to-think-otherwise. "Whenever ... the word 'can' occurs in conjunction with the pregnant use of 'think'," Husserl comments (p. 446), "there is a reference, not to a subjective necessity, i.e., to the subjective incapacity-to-represent things otherwise, but to the objectively-ideal necessity of an inability-to-be-otherwise. This is by its essence such as to be given in our consciousness of *apodictic self-evidence*." (1) here is intended to state that *e* is just such an act or state.

To (1) a clause can be added specifying the *object* of *e*, or what *e* is "about." Thus if *e* is an <Evid> and has as its "object" a statement to the effect that an object *x* has such and such a property, the property is in effect the very "essence" of *x* — a property that *x* must be thought to have, one without which *x* cannot be thought to exist. The use of (1) helps to capture much of what Husserl wishes to say about essence, laws of essence, species, and the like.

As previously, also, let us let

(2) 'x Under_e a'

express that in the intentional act or state e the object x is taken under the one-place predicate-description a. 'Under' here may in fact be defined in terms of the basic pragmatical predicate 'Acpt' for *acceptance* and the semantical one 'Des' for *designation*.[6] Thus we may let

'x Under_e a' abbreviate '(PredConOne a . <Acpt> e . (Eb)(Ec)(b Des x . $c = (a^\frown b)$. e $\text{Of}_{\text{Object}}$ c))'.

a here is to be a one-place predicate-constant of the language and b is some result of *concatenating* a with some designator of x so as to form a sentence. If e is an intentional act of accepting such a sentence, then c is taken under e in that act. Undering, so to speak, is clearly a fundamental kind of act in all language-use.

Apodictic self-evidence, Husserl points out, "is correlated with a definite pure law in each case. It is ... obvious that objective necessity is as such tantamount to a being *that rests on an objective law*.... The inability-to-exist-by-itself of a dependent part points therefore to a *law of essence*,.... *Dependent objects are objects belonging to such pure Species as are governed by a law of essence to the effect that they only exist (if at all) as parts of more inclusive wholes of a certain appropriate Species*."

Let

'Sp x' and 'Gen x'

express now that x is a Boolean sum or collective of entities constituting, respectively a pure Species or a pure Genus in the sense Husserl intends. And let

'Law a'

express that a is a "law" of the appropriate kind (§11, pp. 455ff). These notions are quite fundamental to Husserl's discussion, and we could not get very far into his theory without them. Let also

'Obj x'

express that x is an "object" in just the sense Husserl intends. We may then define

'CollObj x' as '((Ey)(Obj y . y P_H x) . (z)((Unit z . z P_H x) $\supset$ (Ew)(Obj w . w P_H x . z P_H w)))', to express that *x* is a *collective of objects*, i.e., any entity having at least one object as a part and such that all unit parts of it are parts of some object that is a part of it. Every object consists of the sum of its unit parts, so that *x* here can contain only units that are parts of objects.

All species and genera are collectives of objects, and to say that an object is a *member* of a species or genus is now merely to say that it is a part of one. Thus

'*x* ε *y*' could abbreviate '(Obj *x* . (Sp *y* v Gen *y*) . *x* P_H *y*)'.

What now is a "whole"? Let us let

'W_a *x*' abbreviate '(E*e*)(<Evid> *e* . (E*y*)((Sp *y* v Gen *y*) . *a* Des *y*) . ((Obj *x* . *x* $Under_e$ *a*) v (CollObj *x* . (*z*)(*z* ε *x* ⊃ *z* $Under_e$ *a*))))'.

A whole, relative to some mode of description *a*, is any object *x* taken under *a* in some evidencing-act *e*, or any collective of objects every member of which is taken under *a* in *e*, where *a* itself designates either a species or a genus. The object *x* is then supposed to "depend" upon *a*, or to be "founded" upon or to "need supplementation" by *a*.

Let us now turn to the key relations of *dependence*. There are several of them, each determined by what is taken to depend on what. Let us first define

'*x* RelDep *y*' as '(((Sp *x* . Sp *y*) v (Gen *x* . Gen *y*) v (Sp *x* . Gen *y*)) . (E*a*)(E*b*)(E*c*)(E*e*)(<Evid> *e* . *a* Des *x* . *b* Des *y* . Law *c* . *c* = ⌜(*z*)((*a*⌒*z*) ⊃ (*b*⌒*z*))⌝ . *e* Of_{Object}*c*)))'.

This definition is intended to capture — perhaps not altogether satisfactorily — Husserl's comments (p. 406) that "a content *A* is relatively dependent in regard to a content *B* ... if a pure law, rooted in the peculiar character of the kinds of content in question, ensures that a content of the pure Genus *A* has an *a priori* incapacity to exist except in, or as associated with, other contents ... determined by *B*." The pure law here is the *c*, which in effect states that all *A*'s are *B*'s, and this with apodictic self-evidence.

Clearly then

'*x* RelIndep *y*' may abbreviate '((Sp *x* . Sp *y*) v (Gen *x* . Gen *y*) v (Sp *x* . Gen *y*)) . ~ *x* RelDep *y*)',

expressing that *x* is relatively independent of *y* in the absence of a law of dependency. Let also ‘*x* ObjDep *a*’ abbreviate ‘(Obj *x* . W_a *x*)’,

and

‘*x* CollDep *a*’ abbreviate ‘(CollObj *x* . (E*y*)((Sp *y* v Gen *y*) . *a* Des *y*) . (*z*)(*z* ε *y* ⊃ *y* ObjDep *a*)))’,

expressing respectively that *x* is an object dependent on *a*, and that *x* is a (generic) collective every member of which is an object dependent on *a*. Also we may let

‘*x* DepP *y*’ abbreviate ‘(*x* P_H *y* . *x* RelDep *y*)’

and

‘*x* IndepP *y*’ abbreviate ‘(*x* P_H *y* . ~ *x* RelDep *y* . ((Sp *x* Sp *y*) v (Gen *x* . Gen *y*) v (Sp *x* . Gen *y*)))’.

These provide us straightforwardly with the relations of being a dependent and independent part of.

With these definitions in mind, we could attempt to state Husserl's six "propositions" in §14. Let us refrain, however, from so doing. The full content of these is not easy to fathom and they are not above ambiguity. And, anyhow, much of what they say involves the category-mistake pointed out above. As Findlay himself has remarked (p. 22): "Husserl does not consider the possibility that it may be an ontological mistake to apply the notion of part to a non-independent moment, since a true part must arguably belong to the same category as the whole it helps to form (a page and a book) whereas a 'moment', in virtue of its non-independence, is arguably not of the same category as the whole into which it enters. There is something misleading about treating the shape of a book as a part of it...."

No doubt much of the foregoing material may be improved upon. Enough has been shown, however, to see how the root notions of Husserl's *Investigation III* may be handled within the event-theoretic framework, with the addition, of course, of key Husserlian notions such as '<Evid>', 'Law', and so on. And governing all of these notions, meaning-postulates must be given. Such postulates would "interanimate" most intimately with the six propositions, so intimately in fact that these latter are virtually unintelligible without them. Even so, the tentative definitions given above may be helpful to others who wish to probe the foundations of Husserl's theory more deeply than has been done here.

What now, we might ask, is of permanent significance in Husserl's *Investigation III*? It contains little analysis of any specific part-whole relation as such, in the sense of laying down basic properties of it, defining derivative ideas in terms of it, and so on. In these respects, the subsequent work of Leśniewski represents a very considerable advance. Husserl's interest is mainly in relations of dependence, as we have seen. His discussion of this is rather obscure, however, as we have noted, and cries out for a "clean" formulation in modern terms. For such a purpose, it would seem unwise to presuppose a theory of genera and species, nor a theory of <Evid>, which is too intimately dependent upon a very special theory of knowledge. What we want to get at is the full *structure* of a mereology independently of all this; there will then be ample time to interrelate it with whatever theory of knowledge one wishes. Also, Husserl has no mereological "units" other than "objects." And, strictly, he has no notion of a "collective" in any clearly articulated sense. Without Boolean sums we are totally in the dark as to what a "complex" object is supposed to be. Even so, Husserl must surely be credited with having gotten the ball rolling, so to speak, and there can be no doubt of the considerable influence his work exerted on Leśniewski.

According to Findlay, "Husserl's treatment [of dependence] can ... be defended on analogical grounds: 'moments' are not parts, but in certain respects they behave like parts, and the identities and differences of their logical behavior is what the *Investigation III* attempts to determine," Strictly, Husserl should have spelled out these identities and differences in detail by paying more attention to the multiplicity and character of the *relations* stipulating the dependences desired. The modern logic of relations was available at the time to have helped him do this. By 1900, when Husserl's work first appeared, both Frege and Peirce had done much of their work, and Schröder's famous and much-used text-book[7] had already appeared in 1895. In fact the full logic of relations was already in essentially its modern form in Peirce's great paper of 1885.[8] It seems *retardataire* for Husserl continually to speak of genera and species and of "syllogistics" and syllogistic inference, when the full riches of modern logic were virtually at hand.

Let us now return to Mulligan's discussion, which tends to upgrade Husserl at Leśniewski's expense. For him "constituency and dependence" relations are needed fundamentally for the analysis of linguistic structure. Unfortunately he does not spell out for us the exact behavior of either kind of relation. The question arises as to whether the kinds of "dependency" relations he uses are in fact of Husserlian type. Or can they all be handled in clear-cut way within a Leśniewskian kind of scheme — with of course suitable extensions in the manner of event-logic? We should not force too much out of the little words 'part' and 'whole' where and when more accurate terminology is available.

Let us recall that the notion 'Law' is fundamental to Husserl's discussion of dependency. Instead of handling "dependence relations" mereologically, let us seek to handle them instead in terms of the particular laws determining the particular kinds of dependency needed in a given context of inquiry. Such a procedure would be more specific — more "scientific," if you like — and would place dependencies right where they should be, in the laws stating them, these laws being expressed in terms of the relations involved.

Mulligan's discussion centers around an analysis of "the event which consists in John promising Mary that he wouldn't ever see Susan again. One of the parts of this event," we are told, "is John's utterance of the words

(3) I promise that I will never see Susan again.

This utterance consists of meaningful signs. And we can distinguish between the whole utterance and its constituent phonemes and also between the whole utterance and its syntactic parts." Mulligan claims that several "dependence relations" are needed for analyzing (3). "Firstly and most obviously: a relation of dependence at the level of pragmatics: an action of promising cannot occur without a use of the signs 'I promise ...' (or their equivalents in other languages) .... Secondly, ..., we can note that the event which is a promising always marks the beginning of two *states*, themselves mutually dependent, of claim and obligation." Also "there can be no promising event unless we can also identify that kind of interconnection of syntactic parts which we call a proposition or 'that'-clause.... Another necessary constituent of this complex is a certain sort of mental state or mental event, often referred to as the speaker's sincerity/intention." In all this, it should be noted, there is talk of dependency and interconnection, but not strictly of "parts" or "wholes."

In the author's *Semiotics and Linguistic Structure* (pp. 237-238) an analysis of sentences of the general form (3) was suggested, which will be of help here. Let us recall this analysis and then note some of the "dependency" relations required for it.

Let us first note that an appropriate logical form for (3) is, with some simplification, merely

(3′) '(Ee)(Ea)(I Agent e . <Promise> e . e During now . e $\text{That}_{\text{Content}}$ a . '~ (Ee_1)(I Agent e_1 . <See> e_1 . now $\text{Before}_{\text{Time}}$ e_1 . e_1 $\text{Of}_{\text{Object}}$ Susan . (Ee_2) e_1 Again e_2, {e_3 ∋ (I Agent e_3 . <See> e_3 . e_3 $\text{Of}_{\text{Object}}$ Susan)})' a)'.

Agent here is the relation of being the (or a) agent of, During, $\text{Before}_{\text{Time}}$, and Again are suitable temporal relations, $\text{That}_{\text{Content}}$ is the that-relation of content, $\text{Of}_{\text{Object}}$ is the of-relation of being the object of, and {e_3 ∋ --e_3--} is the virtual class of all events e_3 (states, acts, processes, etc.) satisfying the sentential form '--e_3--'. 'I' and 'now' are deictic expressions for the speaker and for present moment, respectively.

(3′), it should be observed, is merely a logical form for (3). But no utterance of it is, as such, a promising. The promising takes place rather *by means of* such an utterance, provided the utterance is "sincere" or "truly intended", or something of the kind. Let '<Utt>' be the event-descriptive predicate for acts of utterance, and let 'Γb' express that b is an inscription of the very form (3′). For me to make the promise intended here, is then captured by the form

(3″) '(Ee_4)(Ea)(Eb)(Ee)(I Agent e_4 . <Promise> e_4 . e_4 During now . e_4 $\text{That}_{\text{Content}}$ a . '~ (Ee_1)(I Agent e_1 . <See> e_1 . now $\text{Before}_{\text{Time}}$ e_1 . e_1 $\text{Of}_{\text{Object}}$ Susan . (Ee_2)e_1 Again e_2, {e_3 ∋ (I Agent e_3 . <See> e_3 . e_3 $\text{Of}_{\text{Object}}$ Susan)}' a . Γ b . <Utt> e . I Agent e . e During now . e $\text{Of}_{\text{Object}}$ b . e By_{Means} e . e High Less-Sincere-Than,' <Utt>')'.

High here is the relation of being placed high in the scale of the comparative relation of being Less-Sincere-Than as confined to (human) utterings.

At first glance (3′) and (3″) might appear complex. Note how simple they are actually, however, with each structural component of (3) merely spelled out explicitly, the result being merely a conjunction with some occasional quantifiers — all within a first-order metalogic.[9] The spelling out of detail here is the important item, as essential in linguistic study as in mathematics or physics.

(3″) can be taken as a logical form for a so-called *performative utterance* of the kind (3). Austin himself insisted, it will be recalled[10], that performative utterances "do not themselves report facts [whatever *they* are] and are not themselves true or false." In a way he is right, b being an occasion sentence, not an eternal one. Occasion sentences, it will be recalled, are true only relative to the speaker (and perhaps act of utterance).

Nonetheless Austin's account is too simple, for sentences of the form (3″) can be true or false in a quite straightforward sense. Either my promise takes place or it does not. Just what the *criterion* for its taking place is, we are not told. "The words have to be said in the appropriate circumstances [p. 223].... But one thing we must not suppose is that what is needed in addition to the saying of the words in such cases is the performance of some internal spiritual act, of which the words are then to be the report." *No, not the report, but the means.* "It is better . . . to stick to the old saying that our word is our bond." *Yes, but our bond is more than our word.*[11]

In addition to the event-descriptive predicates note the many relations needed in (3″): the relation for agency, the three temporal relations, and so on. Are all of these "dependency" relations? In one sense, yes — in the sense that they are all governed by law, i.e., by the meaning-postulates concerning them. All non-logical relations are dependency-relations in this sense. And also in another sense: if e is said to take place now, i.e., to bear the relation During to the present moment, e depends upon the present moment for its temporal location. If I am the agent of e, e depends upon me for its character as something done by me. In general, if x bears R to y, x may be said to depend R-wise upon y, and y to depend converse-R-wise upon x. Of course, this broad sense of dependency is not the one of Husserl, which requires additional factors. If at any point such factors are needed, they may be brought in.

It looks as though Husserl, and Mulligan in his wake, is confusing dependency with relationality. Relations are indispensable in linguistics as elsewhere. And we must construe relations severally, not collectively as Husserl and Mulligan do, symbolizing each one and allowing it to do the work assigned to it. We must not lump vast congeries of relations together under some grand rubic. The analysis of the language of mathematics, as given, say in *Principia Mathematica*, could not have succeeded if Whitehead and Russel had not developed an extensive symbolism for various kinds of relations and then shown in detail how they are needed, over and over again, in the symbolization of mathematical notions. The same is true of linguistics, with the difference that the latter is vastly more complicated. The enumeration of the kinds of relations needed for it has just begun, although before too long we may well have the results of a first approximation.

It is much to Mulligan's credit that he calls attention to the need for relations in the analysis of linguistic structure. In particular, relations for various syntactical (based on concatenation), semantical (based on Des and Under), and pragmatical notions are needed. Metalogic — logical semiotics — thus becomes an essential ingredient of language study, each natural language containing its own metalanguages (according to Zellig Harris). It is also good for Mulligan to observe that "in most languages speakers distinguish, in their speaking and understanding, between (at least) enduring substances, punctual events, extended processes, and enduring states." He claims, further, however, "that these distinctions all rest on relations of constituency and dependence: events, procession and states on substances; the beginnings and endings of processes on events; any state or process can be fragmented into its constituent phrases or portions; substances have other substances and may also have events and processes as parts." And so on. Are there, then, no events, processes, or states not "depending" on substances? *Pace*, then, all process philosophy of Whiteheadian type. Are not the middles of processes also events? Are not all processes events? If a substance undergoes an event, is it not better to say that that event is *of* that substance, or an event occurring to it, rather than a part of it? Again, Mulligan is stretching 'part' beyond its normal limits. Mereology has an important enough role to play as is; it should not be forced to usurp the roles played by the general theory of relations.

Are all "constituency" relations to be handled mereologically? By no means; each type should be given its own vocabulary. Thus, contrary to Mulligan, the letter 't' is not strictly a part of the word 'the'; rather it is a *concatenate* of it. No one would dream of calling 3 a "part" of (7 ÷ 3), say; no, it is the denominator or divisor. Nor is 7 a "part" of the number (7 + 3). Nor is an utterance of a sentence of the form (3) a "part" of a promising. No, it is rather the (or a) means by which the promising takes place. And so on and on.

One final comment, concerning the "claim" and "obligation" involved in an act of promising. In what sense should these be regarded as "parts" of the promising? As temporal parts, perhaps, the two being "mutually dependent", as Mulligan notes. In what sense? we may ask. Perhaps in the sense stated, that two such "states" always begin any act of promising. That this is the case could then be laid down in a meaning-postulate.

Mulligan is surely correct in claiming that the "apparatus of set-theory" is both "alien and descriptively inadequate" for the purposes of founding a viable mereology and/or linguistic theory. It would have been interesting to have been given his reasons. In V above we have seen that set-theory is in a way even alien to mathematics, and in X that it contains so many internal difficulties as to be unacceptable outside of mathematics. And, of course, it is "descriptively" inadequate outside of mathematics, for any purpose whatsoever, without the addition of a very considerable supply of non-logical constants.

CHAPTER FIFTEEN

On compound individuals and the languages of science

> "It is evident that scientists and philosophers can help each other. For the scientist sometimes wants a new idea, and the philosopher is enlightened as to meanings by the study of scientific consequences." Whitehead

In a recent paper[1], the eminent chemist Joseph Earley has pleaded in effect for an articulated metaphysics providing for the notion of a compount individual. And well he might, for one must search far and wide for such, all done up with the necessary refinement of detail we now expect in any original metaphysical writing. If we are content to let our metaphysics just revolve around the views of some great philosopher of the past or near-past, we ordinarily seek no greater precision — or lack of it — than he did. And many metaphysicians do tend to rest content with the metaphysics of the past. When I was an undergraduate at Harvard, I remember asking Professor Demos's permission to write a senior honors-paper on Whitehead. "No," was his answer, "the philosopher you write about must be *well-dead*." What a pity, I thought; Professor Demos does not realize that metaphysics is a subject very much alive, and that the last word about it has not been said. And so it is with the notion of the compound individual. Not only has the last word about it not been said; we scarcely have a reliable first word to help get us started on the right track. And such first words as we do have seem to have been largely neglected, i.e., those provided by mereology or the calculus of individuals.

Let us recall the extended mereology — or *mutatis mutandis*, homology — of V, which, if we disregard for the moment the ordinal individuals, is a *one-sorted* theory, and the basic part-whole relation P has the cardinal entities as its relata. For purposes of application to the language of any branch of empirical science, however, it is usually convenient to subdivide the basic entities into many different sorts. As Earley notes: "Macroscopic entities of everyday experience (including ourselves) are made up of microscopic and sub-microscopic components: brain-cells, molecules, atoms, hadrons, quarks, ..." Do we wish this list to be arranged in a descending order, so that an entity of a given kind is a *part* of an entity of a kind mentioned earlier in the list? No, this would be much too simple. The various kinds of entities any science treats of bear very complex relations to each other — as indeed do the entities of any one kind. And what, anyhow, is a kind, and how do the individuals of a kind interrelate to constitute a kind?

Following the eminent logician Hao Wang[2], let us speak of a *many-sorted* theory as follows. "Certain axiomatic systems involve more than one category of fundamental objects; for example, points, lines, and planes in geometry; individuals, classes of individuals, etc., in the theory of types. It is natural to use variables of different kinds with their ranges respectively restricted to different categories of objects, ... As an alternative way, we may also formulate a system involving several categories of fundamental objects by using merely one kind of variables which have the sum of all the categories as their range of values. The simplest way to bring in the distinction of categories is to introduce n one-place predicates 'S_1', 'S_2', ..., 'S_n', [quotes added] such that x belongs to the i-th category if and only if $S_i x$." Let us use Wang's notion here in reflecting upon a *many-sorted mereology*.

The theory of V made much of unit entities, and these in effect constituted a sort. The units are the *indivisibilia* of the theory. From a general metaphysical point of view, the assumption of such entities might be thought dubious. Perhaps there are no indivisible elements; perhaps the most fundamental objects are *continua* of some kind, such as waves or forces or pulses of energy. In any case, it might be thought, a mereology adequate for the logical foundations of any science should not rule out the possibility of their being no *indivisibilia* at all. Let us therefore drop the units of the theory of V, and in their place allow for many sorts of objects. Let 'S_1', ..., 'S_n' be the primitive predicates for the many sorts, their precise character being studied in the given science.

The various Boolean notions are still available in the mereology without units. Thus, given any entity or "individual" of the sort S_i, Boolean sums of such individuals are available. Such sums are in effect "collectives" of individuals of the sort. Let us define

'$\mathrm{CollS}_i\, x$' as '$((\mathrm{E}y)(\mathrm{S}_i\, y \,.\, y \,\mathrm{P}\, x) \,.\, (y)(y \,\mathrm{P}\, x \supset (\mathrm{E}z)(\mathrm{E}w)(\mathrm{S}_i\, z \,.\, z \,\mathrm{P}\, x \,.\, w \,\mathrm{P}\, z \,.\, w \,\mathrm{P}\, y)))$',

to express that x is a sum of individuals of the sort S_i . This obtains just where some S_i is a part of x, every part of x having a part in common with some S_i that is a part of x. The definiendum here also could be written as

'$\mathrm{CompS}_i\mathrm{Ind}\, x$',

expressing that x is a compound individual of the sort S_i. Even the whole sort is a compound individual in the sense that that individual is the *fusion* (in the familiar sense of the calculus of individuals) of that sort, where

'$\mathrm{FuS}_i\, x$' is defined as '$((y)(\mathrm{S}_i\, y \supset y \,\mathrm{P}\, x) \,.\, (y)(y \,\mathrm{P}\, x \supset (\mathrm{E}z)(\mathrm{E}w)(\mathrm{S}_i\, z \,.\, z \,\mathrm{P}\, x \,.\, w \,\mathrm{P}\, z \,.\, w \,\mathrm{P}\, y)))$'.

Clearly for each S_i it is assumed that

$\vdash (\mathrm{E}x)\mathrm{S}_i\, x$;

otherwise there would be not point of introducing 'S_i' in the first place. It then follows of course that

$\vdash (x)(\mathrm{FuS}_i\, x \supset \mathrm{CollS}_i\, x)$.

Although no "units" have been needed in order to define these — and indeed other — Boolean notions, we could take S_1 as the units if desired. In any case with or without units a many-sorted mereology gives us a way of aggregating individuals within a given sort so as form compound individuals of that sort. It is not clear that any other clear-cut method of forming compound individuals has ever been put forward. To be sure, for centuries there has been loose talk, both in metaphysics and in the sciences, about parts and wholes, but with almost no background theory to provide for it. We should then, no doubt, welcome many-sorted mereology with open arms as giving us a foundation for something we thought we had had all along.

Concerning each S_i, of course, a great deal has to be said, and some of this will be in terms of individuals of the other sorts. We could not get very far in a theory of physical atoms without a good deal of prior talk about protons, electrons, and so on. And in all this, of course, a great deal must be made of the *relations* needed. Relations are forthcoming in full supply as values for variables in ordinal mereology, so that the problem for each S_i is to pick out such fundamental or primitive relations as are needed to characterize the S_i-individuals and their behavior relative to one another and relative to individuals of the other sorts.

All the compound individuals or collectives admitted thus far have been of entities of the same sort. Nothing rules out the admission of compounds of entities of different sorts, however, which might be useful for some purposes. Let $i \neq j$, so that S_i and S_j are mutually exclusive sorts. Then we may let

'$CompS_iS_jInd\ x$' abbreviate '$((Ey)((S_i\,y \vee S_j\,y) \,.\, y\,P\,x) \,.\, (y)(y\,P\,x \supset (Ez)(Ew)(((S_i\,z \vee S_j\,z) \,.\, z\,P\,x \,.\, w\,P\,z \,.\, w\,P\,y)))$',

to express that x is a compound individual of individuals of the two sorts. And similarly for compounds of individuals of any three sorts, and so on.

In a many-sorted, ordinal mereology we have the wherewithal now to talk of S_i-individuals and of compounds or collectives of such in the same breath, so to speak, and on the basis of a unified ontology. We are in a position to interrelate the behavior of an individual with that of a compound in whatever way the science at hand requires. Consider biology, for example, and the problem of what the units of evolutionary selections are. As Earley points our, Dawkins takes the gene as fundamental, Lewontin the individual, Wynne-Edwards the species. But all of these factors play some role or other, surely, as E.O. Wilson has emphasized. The problem of determining the precise role of these and other factors is not a metaphysical problem but one internal to evolutionary biology itself; it is a matter of specifying the appropriate relations each bears to each.

A valuable, but much neglected, beginning toward formulating some of the notions needed in any language that talks of evolution, is to be found in Appendix B to J.H. Woodger's Tarner Lectures.[3] There, it will be recalled, the notion of an "evolutionary relation" was introduced in terms of prior notions of a "taxonomic set," a "P-line," and a relation of "joining" P-lines. P here is the relation of parent to offspring, a most fundamental one surely in all evolutionary study. A taxonomic set is any species, genus, family, sub-genus, and so on. A P-line — finite, with a beginning and an end — is any sub-relation of P such that there is a one-one relation R whose final term (or relatum) is reached in a finite number of R-steps from its first term (or referent).[4] For the relation of joining, we may let '$J(R,X,Y)$' express that R is a line, X and Y are taxonomic sets, and the beginner of R is a member of X but not of Y, and the end term of R is a member of Y but not of X.

A given taxonomic set X stands in an "evolutionary relation" E to a taxonomic set Y now, just where there is some P-line R such that $J(R,X,Y)$. E would seem to be the fundamental relation of any theory of evolution, no matter what its details might be or what units of selection are taken to be the fundamental ones. The problem of formulating a theory of evolution is now one of laying down postulationally the specific properties E is supposed to have. Such postulates would help to make precise the role that genes, individuals, species, and so on, all play in evolutionary development. Woodger suggests tentatively several such postulates, by way of a first attempt to supply such.

Note that Woodger's work is couched in the language of set-theory. With only slight changes, the essentials of it may be reformulated in mereological terms. In place of a taxonomic set, let us speak of a taxonomic collective, and all talk of relations may be carried out in terms of ordinal individuals and Boolean sums of them, as in V. A "member" of a taxonomic compound S_i is then any S_i that is a part of it. Woodger's J in turn can be handled then as a Boolean sum of all triples (r_2 ord (x ord y)), where x and y are taxonomic collectives and r_2 an appropriate sum of ordinal individuals the ordinands of which are collectives. And similarly, as needed, for the relation E. In this way we see that evolutionary theory can be handled as well in mereology as in the more usual set-theoretic mold. (Woodger himself, incidentally, who made much of mereology in his earlier work on embryology and cytology, would very much have welcomed the extensions of it that allow the simple treatment of relations needed here.)

Woodger's work has opened the door to a kind of mathematical-logical theory of evolution. The work of M.B. Williams, Aristid Lindenmayer, and others, carries on where Woodger left off. For present purposes, let us reflect very briefly on the paper by Williams given at the Salzburg Congress in July, 1983.[5] She discusses "the two most central philosophical problems: (1) what is the relationship of the various entities involved in selection (e.g. gene, organism, species) to the fundamental laws of the theory?, and (2) how are the 'levels of selection' related to the fundamental laws of the theory?" How good to focus immediately on the *laws* of the theory — without which we have no theory — and upon the *relations* needed to frame those laws. "Biologists have frequently assumed that there is *one* fundamental entity (*the* unit of selection) such that the effects of selection on any other entity is the result of its effects on this fundamental entity." Williams thinks this assumption is an error. She goes on to speak of an "axiomatization" of the theory and of various "models" of it, where it would perhaps have been more helpful to have developed a unified theory in which all the models are incorporated. Each can then be studied in its own terms and interrelated with the others. Situations in which the "group" or "deme" or "generation deme pool" (Wade), the "altrustic trait" or "kin-group" (Hamilton), the "gene" or "gene-pool" (Fisher), "chromosome", and so on, were relevant could all be handled separately, as well as jointly, within a unified theory. Every kind of selection-process can then be studied in its own terms and not confused with those appropriate for another. The terminology used for any one is "relative to the evolutionary problem being investigated," as surely it should be.

In speaking of evolutionary processes, 'process' should be taken seriously. Processes should not be confused with the entities undergoing process or change or whatever. For many purposes, it is convenient to recognize events in general, including acts, states, processes, changes, and so on, as values for variables. Let

'<BProc> *e*'

express that *e* is a biological process, where '<BProc>' is an event-descriptive predicate. Individuals, genes, chromosomes, and so on, all enter into biological processes in complex ways. The recognition of many types of such processes as fundamental entities of the theory would seem a useful adjunct, and enables us to express much that cannot be expressed otherwise — i.e., unless times themselves are admitted. (Woodger's own method was to include a special temporal relation T.) Among these the *selection*-process would be of major importance. Let

(1) '<Sel> *e*'

express that *e* is such a process. What entity selects what, when, and where along a given evolutionary line? The use of (1) would seem essential for any precise terminology in which such matters may be discussed.

Let us reflect for a moment now on Werner Leinfellner's work on the evolution of intelligence.[6] For this, fundamental use of processes must be made, even of mental processes, which cannot be "reduced" to brain-states — in any easy-going way anyhow. "Intelligence" Leinfellner regards as "the degree of efficiency of cooperation of four factors when solving problems." These are, roughly, a "sensorium or information input-device; a memory or storage device for input-output information and successfully played games, a computing or inferring system, and a motoricum or realizing device." The theory of all of these needs a careful looking at. It would seem almost unthinkable that it could be formulated without the linguistic resources of an event-logic. Any theory of "problem solving" is a theory of a kind of mental activity in which declarative sentences (concerning the "options" open) are compared with each other in various ways. Leinfellner makes fundamental use of game-theoretical methods, and these too seem best formulable in act- or event-theoretic terms.[7]

Mention of mental events leads us back to Earley's paper. Especially important for him is "the problem of the compound individual." What, after all, is this problem? Presumably it is this: when is an individual of any sort to be taken as an indivdual *simpliciter* and when as a compound? Earley attributes to Whitehead the view "there is *no* univocal resolution to the problems of the compound individual. No entity is, simply, one. There are no 'elementary' entities: everything is, in some sense, composite. Every one is, somehow, many. There is no *level* of entity that is, univocally, fundamental.... If there is no one *level* (spatial or temporal) of entity that is non-composite, is there some *kind of interrelation of parts* that gives rise to unification that merits 'metaphysical status'?" But for Whitehead, surely, the "actual entities" are univocally fundamental, to say nothing of the eternal objects. Each and every AE is "one," but either prehends or is prehended by other AE's. Each EO is "one" but is prehended by AE's and bears various logical relations with other EO's. Prehension and ingression are Whitehead's wherewithal for providing the kind of unification of entities that merits metaphysical status.

This, however, is not Earley's view. He contends, rather, that "an entity should be regarded, metaphysically, as *one* entity if, and to the extent that, it interacts with particular percipients as unified." The sun, for example, he states, "is a structure, ... one that necessarily involves net production of entropy, and concomitant interaction with the rest of the universe.... The sun has [as] much claim to be considered, metaphysically, as one entity as has any other candidate for the status (any human person, any miniscule space-time region). The sun, as a unified whole, makes a difference for the rest of the world in virtue of its output (which depends on its interpersonal processes): that output is sufficiently stable that other entities can adjust themselves to its existence...." But cannot similar contention be made of *every physico-chemical entity* whatsoever? All interact with "percipients" as unified; all make a difference with the rest of the world; other entities adjust themselves to their existence. And what, anyhow, is a "percipient," metaphysically regarded? Merely one of the relata of the relations expressing the interactions.

In speaking of wholes and parts, much would seem to depend upon the context of discussion and the *way* in which given entities are spoken of. In a sentence of the form

$$`(x)(S_{(i+1)}\, x \supset \text{--}x\text{--})\text{'},$$

the $S_{(i+1)}$-individuals are spoken of as wholes. But suppose now that every $S_{(i+1)}$ individual is a compound of S_i-individuals. In the sentence

$$`(x)(\mathrm{CompS}_i\mathrm{Ind}\, x \supset \text{--}x\text{--})\text{'},$$

the $S_{(i+1)}$-individuals are spoken of as compound S_i-individuals. Both modes of locutions are useful, but one kind may be appropriate in one context, the other in another. If we wish to speak in both ways in one and the same sentence, we may bring in the Under relation used previously.[8] Where

$$`x\ \mathrm{Under}_e\ a\text{'}$$

expresses that *x* is taken under the one-place predicate *a* in the intentional act *e*, we might have both that

$$(x\ \mathrm{Under}_e\ `S_{(i+1)}\text{'} \,.\, x\ \mathrm{Under}_{e'}\ `\mathrm{CompS}_i\mathrm{Ind}\text{'}).$$

x may be taken under both descriptions without contradiction in separate intentional acts *e* and *e'*.

Another interesting use of the theory of mereological sums is in the handling of biological species, concerning which there has been much recent discussion. In several important papers[9], David Hull has been urging that biological species be regarded as "spatio-temporally localized individuals," rather than as classes, expecially for the purposes of a theory of evolution. Hull does not quite say so, but such "individuals" are no common run-of-the-mill things, but "compound" individuals in essentially the sense discussed here. The details needed to substantiate this contention remain to be worked out, as do the details for evolutionary theory generally. This latter, as Hull remarks, "is currently undergoing a period of rapid development. But in the process several problems have cropped up that are proving to be infuriatingly difficult to resolve.... In the present case, I think two assumptions are at fault: (a) the view that genes and organisms are "individuals" while populations and species are "classes," and (b) our traditional way of organizing phenomena into a hierarchy of genes, cells, organisms, kinship groups, populations, species, and ecosystems or communities." If species can be handled as compound individuals or mereological sums, then so

(2) (<Saw> e . $(x \cup y)$ Agent e . e $\text{Of}_{\text{Object}}$ (the tree) . e $\text{With}_{\text{Instrument}}$ (the two-man saw)),

where we would not say that x alone is the agent of such a sawing or that y is. No, the agent is the mereological sum (or team) of x and y. Such usage is common, in which agency can be correctly ascribed to a collective of some kind, but not to the individuals comprising it. In describing such agency, however a clause is usually needed giving the instrumentality or means involved, or some 'how'-clause giving the manner of the action. No difficulty should arise in ascribing agency and the like to collectives provided the means or instruments or whatever are described. A further description can also be given as to just what it is that the individual person does when he participates in the group action. In the sawing example, x and y face each other, one pulls the saw while the other pushes, and so on. The exact logic of joint actions, doings, agency, and the like, seems not to have been studied at much depth, as already noted. It is good that DeGeorge has called attention to it and emphasized its need for social theory. And of course any description of how individuals act in group action would run the risk of being "inaccurate if it ignores the structure and organization that informed their actions, and the fact that they acted together."[12]

Let us turn now again to many-sorted mereology, and note how well it accords with Abner Shimony's plea for pluralism, in his contribution to the Lund Workshop on Parts and Wholes.[13] "Epistemological and methodological problems concerning parts and wholes should also be treated holistically, with attention to the great differences among types of composite systems." Shimony reflects upon four types of systems: "(i) composite microphysical systems in pure quantum states; (ii) microphysical systems undergoing phase transitions; (iii) organic cells considered as cybernetic systems; (iv) typical human societies." He shows convincingly the many-sortedness of each of these systems. The problem of formulating any such system mereologically is left open, but no doubt could be carried out with a suitable choice of primitive S_i's and relations concerning them and inter-relating entities of one type with those of another.

As over and against scientific pluralism, there is always the haunting problem of the unity of science — or at least of the unity of scientific language. It is here, perhaps more than anywhere else, that metaphysics steps in. One of the great functions of metaphysics, throughout the ages, has been unificatory. Piece-meal systems are splendid for special scientific purposes, but never quite suffice for an over-arching metaphysical scheme. The question arises as to whether a many-sorted mereological system can be envisaged to include the various special scientific ones. Envisagement is far from explicit formulation, and this latter is not easy to come by even in the special sciences. It is commonly assumed, however, in methodology and the study of the logic of the sciences, that such formulation is possible; and there is surely abundant evidence in support of it. Well, if so, why can we not then extend our many-sorted mereology to embrace the many-sorted theories required for the special sciences? The broad outline of such a theory is clear enough — merely let our sorts now be handled by a list S_1, ..., S_n, T_1, ..., T_m, U_1, ..., U_k, and so on, together with the requisite primitive relations. The envisagement of such a theory is no more or less speculative than the envisagement of a unified theory for any special science; it is merely a kind of "summation" of such. The underlying ontology of any metaphysical, many-sorted mereology of this kind would then be wholly determined by the ontologies of the special sciences. And of course there would be much overlap here, and some "reduction" would be appropriate, both as concerns the basic sorts and the relations required.

Once all this is said and done, it might be urged, the real problems of metaphysics have scarcely been touched. Metaphysical "holism" is the view that all the sorts of entities required in the sciences, and indeed in the study of values, the arts, history, and so on, are all "reducible" to some one basic sort or "stuff" or whatever. Holism in this strict sense is a view that transcends all the special kinds of study it includes. If metaphysics is possible — as many of us believe it to be — it is an holistic enterprise differing in important ways from more special kinds of study.

Shimony comments that "for many years Whitehead's philosophy of organism and the related monadology of Leibnitz seemed promising to me, but perhaps their obscurities outweigh their promise." And, again, in his sequel[14], Shimony comments that "the program of physicalistic explanation of mental phenomena seems to me hopeless....Whitehead's philosophy of organism offers a possible solution to the problem.... [It provides] a theory which accounts coherently for both the manifestly animate and apparently non-animate sectors of the great chain of being, as the theory of evolution demands. Whitehead's account commands respect in spite of great difficulties in it." Shimony himself has explored deeply some of these difficulties as regards quantum mechanics.[15]

One of the most basic problems in the Whiteheadian scheme concerns the way in which the AO's, the actual occasions, form nexūs, enduring occasions, multiplicities, societites, and the like. What now is a nexus? Recall Whitehead's comments (*Process and Reality*, p. 29) that "actual entities involve each other by reason of their prehensions of each other. There are thus real individual facts of the togetherness of actual entities, which are real, individual, and particular, *in the same sense* [italics added] in which actual entities and the prehensions are real, individual, and particular. Any such particular fact of togetherness among actual entities is called a 'nexus' (plural form is written 'nexūs')...." Again, according to the fourteenth Category of Explanation (p. 35), a nexus is "a set of actual entities [— a *set*, note —] in the unity of the relatedness constituted by their prehensions of each other...." A multiplicity (a "pure disjunction of diverse entities") (p. 36) "consists of many entities, and its unity is constituted by the fact that all its constituent entities severally satisfy at least one condition that no other entity satisfies." A "society" (p. 50) is "a nexus with social order" and an "'enduring object', or 'enduring creature', is a society whose social order has taken the special form of 'personal order'."

Let us not worry here about the characteristics of social and personal order — which may readily be handled relationally — but rather about the ontology status of nexūs in Whitehead's scheme. We cannot identify them with "facts of togetherness" in any strict sense unless such facts themselves are fundamental in the cosmos for Whitehead, which they are not. Nor are they just AO's *simpliciter*, although they do consist no doubt of some AO's taken somehow as "together." They cannot be sets or classes — contrary to the fourteenth Category of Explanation — for these latter are presumably to be handled as EO's. (Whitehead's use of 'set' here is the common, naive one of mathematicians at the turn of the century before any sophisticated use of it came on the market. It should not be taken as a technical term in the present context.) The fact is, it seems, that Whitehead needs a new logico-metapysical notion here, in order to be able to form a nexus from its constituent AO's. A nexus must be "real, individual, and particular, in the same sense in wich actual entities and the prehensions are...." If a nexus were a "set" it would have to be "real, individual, and particular" in a very different sense. The Boolean sums of mereology can provide for these. Nexūs thus become suitable sums of AO's, not just any sums, but those exhibiting a suitable togetherness "by reason of their prehensions of each other." Multiplicities are nexūs of AO's satisfying some given condition. Let

$$(x1\text{--}x\text{--})$$

be the mereological sum of all Boolean units *x* satisfying the condition that --*x*--. The AE's function in Whitehead's system as the Boolean units do in mereology so that (*x*1--*x*--) — for appropriate '--*x*--' — is a multiplicity. Its constituent units satisfy severally the condition that --*x*--, and no other units do — for if they did they would be unit parts of (*x*1--*x*--) and would thus satisfy the condition that --*x*--. Likewise societies and enduring objects may be handled mereologically by bringing in the relations determining the order involved.

It might appear that we are forcing the modern calculus of individuals, or mereology, upon Whitehead's text. But no. It was my good fortune to have been a direct student of Whitehead during his very last year of teaching. The calculus of individuals was being much discussed at that time. The Polish logician Leśniewski had already published his work on the subject in 1927-1931.[16] Independently of this, another rendering of the calculus of individuals was put foreward by Henry Leonard in his doctoral thesis in 1930, a later version of which was published in 1940 by him and Nelson Goodman.[17] I recall mentioning this work to Professor Whitehead during a private conversation at his home. He expressed interest in it and clearly sensed its philosophic importance. My feeling was, moreover, that he thought it would have been useful for him in writing *Process and Reality*. He did not explicitly say this, but merely hinted at it.

The logical scheme underlying the Whiteheadian cosmology can perhaps be best viewed not just in terms of the theory of types, as usually done[18], but as involving a mereology as well. And, in fact, as we have seen in V above, even the theory of types itself can be viewed as contained implicitly in an extended mereology. In fact, the whole Whiteheadian scheme could now be construed as an extended mereology with suitable provision made for the other Categories of Existence. 'Prehension' for Whitehead is a kind of blanket term for all manner of relations, and these are well provided in the scheme here. And of course relations play a central role in modern mathematics, as Whitehead and Russell were the first to show in explicit form. In place of 'prehension', it is often helpful, in reading Whitehead, to ask: what prehension, what "fact of relatedness"? In *Principia Mathematica* the types of relatedness are always well-specified, but not always so in *Process and Reality*. I once asked Whitehead; Why not? His answer was that he came to professional philosophy too late in his life, and that he must leave such quasi-mathematical details to younger scholars. We must thus never forget for a moment that, as Victor Lowe has put it[19], "the only possible answer to the question, [Was Whitehead primarily a] Mathematician or Philosopher? is: At all times, both." And in fact, the two strands of Whitehead's thought were so intimately intertwined all his life, that we cannot understand one without understanding the other.

In his concluding comments at the Lund Workshop, Patrick Suppes noted that "like so many other concepts in science and philosophy, the shimmering beauty of some clear and simple notion is forever appealing and seductive. In fact, in many cases it may remain an ideal always and everywhere, never to be realized but always to be sought after. I am for the delicate dance from parts to wholes and back again. We should not be captured at either end. The dance should go on forever." And indeed it will. In some choreographic contexts it is a solo or a *pas de deux* or *trois* that is appropriate, and in others the full *corps de ballet*.

Another kind of delicate dance, implicit in the foregoing discussion, is the Whiteheadian one between "the 'genetic-functional' interpretation of first principles and the 'mathematical-formal' interpretation."[20] "John Dewey has asked me to decide between [them] ...," Whitehead writes. "But I must decline to make this decision. The beauty of philosophy is its many facets. Our present problem is the fusion of the two interpretations. The historic process of the world, which requires the genetic-functional interpretation also requires for its understanding some insight into those [mathematical-formal] ultimate principles of existence which express the necessary connections [or relations] within the flux." Here too the dance between these two "interpretations" should and will go on forever. The secret is in achieving the appropriate balance to the neglect of neither. At the moment, in metaphysical writing as well as in the theory of parts and wholes, it is the mathematic-formal that is the more neglected. This is a pity, and indeed most contrary to Whitehead's own admonitions as well as actual practice.

It must not be thought the mereological handling of compound individuals is a cure-all for all scientific and philosophic problems concerned with them. Not at all. The mereological treatment merely helps us to view them in clearer focus than heretofore and in terms of an exact framework where previously we had none. Like St. Augustine on time, we think we know what compound individuals are until we being to ask searching questions about them. The point of this paper has been merely to urge that the inclusion of mereology within our underlying logico-metaphysical apparatus is methodologically important and valuable.

CHAPTER SIXTEEN

On Weiss's early papers in logic

"Nec temere, nec timide."

In all logico-philosophical writing we must be careful to distinguish logical from epistemic matters. To be clear about the former is a *sine quá non* for becoming clear about the latter. Likewise a considerable mastery of mathematical technique is needed for the former, without which little progress can be made. A pianist cannot hope to perform Beethoven's Opus 111, say, without years and years of patient, laborious practice, starting with scales and arpeggios and slowly working up to the mastery of delicate matters concerned with pedalling, tone-color, phrasing, and so on. The logician's technique consists of mastery of symbolization and of how signs or symbols relate to whatever it is that is symbolized — in short, of logical semiotics. Very important here is the distinction between the use and mention of expressions and between a language and its metalanguages. Also the logician must have a robust sense of the character of the objects symbolized, lest all manner of will-o-the-wisps lure him into their net unwittingly.

A remarkable fact about technique is that it reveals its secrets only to the initiate. The unwary listener may or may not be able to tell the difference between a performer with a technique adequate for the task at hand and one who lacks it. Sometimes one's aim may be far beyond what one's technique is capable of. The performer's aim may be admirable, but the means for achieving it may not have been sufficiently mastered. One may have an important idea, but not have tarried its "cleansing," in the trenchant phrase of the great dancer Hanya Holm.[1] And after cleansing it must be transformed into what is appropriate in the context. This is all very easy to say, but very difficult to achieve in practice.

Let us examine Paul Weiss's "Eighteen Theses in Logic"[2] of 1957, with an eye to getting at their central content in order to cleanse them and to view them in the light of recent work in semantics, pragmatics, event or process-logic, and the like. We will then be in a position to cast a glance at Weiss's early papers in logic in order to assess them in similar fashion. Weiss is one of the most eminent contemporary metaphysicians, and an assessment of his work in logic is long overdue.

Weiss's first thesis is that "a unique characterization is a predicate, not an attribute," and the discussion centers in part around *14.02 of *Principia Mathematica*. This, it will be recalled, is that

$$E!(\imath x)(\emptyset x) \;.=:\; (Eb):\emptyset x \;.\equiv_x.\; x = b \text{ Df.}$$

Weiss takes this as an assertory "proposition" whereas actually it is merely a definition of notational abbreviation. Thus, strictly, it "asserts" nothing at all. Instead of upon *14.02, Weiss is reflecting upon *14.11, the theorem based on *14.02 with '≡' replacing the '= --- Df' sign. He comments that "if by a "value which satisfies a function" is meant an "existent value", the *Principia Mathematica* [*14.11] asserts nothing but a harmless tautology; but if it is intended to affirm that if there be one and only one value ... [satifying a given function], that value does in fact exist the way horses or tables do, it surely is in error." This contention seems to rest upon a misunderstanding. *14.02 and *14.11 are concerned with the "existence" of *specific* objects, not with horses or tables in general. We must distinguish between

'$(Ex)Hx$' and '$E!(\imath x)Hx$',

expressing, respectively, that there is at least one horse and that there is one and only one horse. *14.11 is in fact a harmless tautology of the system. 'Exists', if used as a technical word, is to apply to the objects taken as values for the variables of that system in some given "interpretation." The logician's use of 'exists' is not to be distinguished from that of the metaphysician. That of the former is to indicate existence internal to the system, so to speak, that of the latter existence, howsoever construed, internal to the metaphysical system.

By way of an example, let us reflect a little on Peirce's horse Cora, say, and let

'Cora_P' abbreviate '$(\imath x)(Hx \;.\; x$ Owned by P. in 1920)'.

In PM-ese, if we assume that the requisite predicates are available, it holds that $E!\text{Cora}_P$, provided that Peirce's horse Cora is taken as one of the values for the variables of the system. If not, not. Cora_P remains, however, in either case, a "real entity, existent in a hurly-burly universe, streaked with contingency." Weiss thinks that any such entity "will be distinct from itself as symbolized in logic or mathematics no matter how precise and inclusive the symbolized form be. A value, a logically proper name, a unique description applicable to a real being, are all distinct from what exists apart from them." But note here the failure to distinguish between use and mention. The "value" here is the entity Cora, which is not the same as 'Cora_P', the *name* of Cora. Names and definite descriptions can, however, be applicable to and uniquely describe real, hurly-burly beings. And it is a mere truism that real beings "are all distinct from what exists apart from them," where 'exists apart from' is presumably equivalent to 'exists as distinct from'.

"In language or logic," Weiss goes on, "predicates, values, etc., are not altogether determinate; outside language or logic their counterparts are attributes — determinate facets of existent beings." The word 'determinate' here is slippery, and anyhow predicates are linguistic expressions for attributes and never to be identified with attributes themselves. Of course things are what they are, and in language we talk about them in certain ways, in ways that the predicates (including relational ones) of the system allow. If some desired predicate is missing, add it to the primitive roster of predicates in order to make the system capable of achieving greater determinateness of expression than before. There is no limit to this kind of extension; and *meta*-logical systems may be formulated even to incorporate the very logic of 'determines' in *whatever sense* Weiss is taking it. There are no upper bounds here to prevent such formulation.

Weiss goes on to point out that the Aristotelian use of 'all' refers to a general, or "actual totality of existents," whereas for the symbolic logician, "'all refers to a universal, a meaning." This way of wording the matter is somewhat misleading. We should consider, rather, 'all' in a context 'all *F*' or 'all *F*'s', where '*F*' is a monadic predicate. Further 'all' is not a word that, in any strict sense, "refers" at all. It is rather syncategorematic, and if it is said to refer, this must be provided by definition.[3] The existential quantifier, within some system — and it has no meaning outside of some system or other — carries the import of what we vaguely express in ordinary language as 'there is (or exists) at least one *x* such that', where 'is' or 'exists' are taken for just the entities that are the values for the variables of that system. The latter are stipulated by a *semantical rule* specifying the objects intended. If no such rule is supplied — or at least in the offing — we have no system in any strict semiotical sense. If the rule specifies that the objects include just "existent" objects, *in whatever sense Weiss intends*, so be it. Or it might also specify the inclusion of "nonexistent" entities, whatever these are — it is up to Weiss to specify here his intent.

Weiss comments that the symbolic logician overlooks that from

(1) 'All (non-existent) gryphons at Yale are men'

we can legitimately derive

(2) 'Some (non-existent) gryphons at Yale are men'.

This is not the case, of course, if the values for the variables of the system include only existent entities and not gryphons. To see what Weiss intends here, let us return to his contention that, for the symbolic logician, "'all' refers to a universal, a meaning," or, equivalently, that "'all' refers to an intellectually distinguished part of an intension or meaning." Aye, here's the rub: to preserve the Aristotelian inference from 'all'- to 'some'-sentences, we must bring in *intensions*. In terms of a clear-cut theory of intensions utilizing the relation Under[4], let us examine Weiss's desired inference of 'Some (non-existent) gryphons at Yale are men' from 'All (non-existent) gryphons at Yale are men'.

Let us let

'x Under_e a'

express metalinguistically, as previously, that x is taken or described under the one-place predicate a in the "intellectual" or intentional act e. Let '$\text{G}\,x$' express that is a gryphon and 'x $\text{At}_{\text{Location}}$ Yale' express that x is *at* the location Yale.[5] The form

(1′) '$(x)(a)((x\ \text{Under}_e\ a\ .\ \text{`}\{y \ni (\text{G}\,y\ .\ y\ \text{At}_{\text{Location}}\ \text{Yale})\}\text{'}\ a) \supset \text{M}\,x)$'

expresses then that in the act e everything taken under the description of its being a member of the virtual class[6] of gryphons at Yale is a man. To justify the Aristotelian inference desired, it must be assumed that at least one entity — perhaps only the null entity — is taken under this description in e, i.e., that

(3) '$(\text{E}x)(\text{E}a)(x\ \text{Under}_e\ a\ .\ \text{`}\{y \ni (\text{G}\,y\ .\ y\ \text{At}_{\text{Location}}\ \text{Yale})\}\text{'}a)$'

holds. From (1′) and (3), it follows by standard logic that

(2′) '$(\text{E}x)(\text{E}a)(x\ \text{Under}_e\ a\ .\ \text{`}\{y \ni (\text{G}\,y\ .\ y\ \text{At}_{\text{Location}}\ \text{Yale})\}\text{'}a\ .\ \text{M}\,x)$'

holds. Weiss's rendition of the Aristotelian form of inference is thus seen to be valid when construed intensionally in terms of the metalogic governing the relation Under, even where there are no gryphons at Yale and even no gryphons at Yale that are men.

The foregoing has been couched in terms of inference, and thus (1′) and (3) have been assumed to be true of *e*. It would be better to contend here only that

$$\ulcorner((1') \,.\, (3)) \supset (2')\urcorner$$

is a logical truth. We can then leave aside assuming either that someone does or does not perform an intellectual act *e* satisfying (1′) and (3).

Weiss's second thesis is that "a "collective" disjunction is distinct from a "distributive" disjunction." Let us recall Aristotle's famous discussion of the sea-fight tomorrow (*De Interpretatione*, 19a 30 ff.): "A sea-fight must either take place tomorrow or not; but it is not necessary that it should take place, neither is it necessary that it should not take place, yet it is necessary that it either should or not should take place tomorrow." (Weiss adds parenthetically that the gist of the problem raised by this passage may be seen "perhaps more evidently" as follows: "Although it is true that I will either die or not die tomorrow, it is not now true that I will die tomorrow, nor is it now true that I will not die tomorrow...." Thus, to begin with, let us speak of truth here rather than of necessity.) Weiss comments that "Aristotle was here distinguishing between '*f*(*x* or not-*x*)' and '*f*(*x*) or *f*(not-*x*)', although he did not provide the logical terms with which to do it." Nor, alas, does Weiss. He comments that "'*f*(*x* or not-*x*)' makes use of a *collective* disjunction: here *x* and the non-*x* are not separated off from one another; they are so separated in 'f(*x*) or f(not-*x*)' which makes use of *distributive* disjunction."

To handle all this let us let

'*e* Occ *t*'

express that the event *e* occurs at or during the time *t*, and let 'today' and 'tomorrow' be deictic expressions for today and tomorrow, respectively. Similarly let

'$\mathrm{Tr}_t\, a$'

express that the sentence *a* is true-at-time-*t* in the sense of a temporalized semantical truth-predicate.[7] We may then give the following logical form to Aristotle's famous statement, construing 'necessary' as 'eternally or timelessly true' (i.e., 'Tr' without the subscript '*t*'):

> '(E*e*)(E*a*)(SeaFight *e* . *a* Des *e* . Tr ⌜(*a* Occ tomorrow v ~ *a* Occ tomorrow)⌝ . ~ ($\mathrm{Tr}_{\mathrm{Today}}$ ⌜*a* Occ tomorrow⌝ v $\mathrm{Tr}_{\mathrm{today}}$ ⌜~ *a* Occ tomorrow⌝))'.

The use of the event-variable '*e*' is needed to handle the sea-fight, 'Occ' to handle its occurrence or lack of it, the temporal expressions for the times involved, 'Tr' (with or without the temporal subscripts) for the two kinds of truth needed, and the usual truth-connectives and quantifiers. 'Des' for designation, the expressional variable '*a*', and the corner-quotes are needed in order to avoid confusion between use and mention. ('*e*' is a variable not a constant, so that 'Tr '*e* Occ tomorrow'', for example, could never hold, '*e* Occ tomorrow' being a sentential form, not a sentence, and 'Tr' being applicable only to sentences.)

Note that in all this there is no need for the distinction between a "collective" and a "distributive" disjunction. Weiss's forms '*f*(*x* or not-*x*)' and '*f*(*x*) or *f*(not-*x*)' are too simple. "The collective disjunction occurs in the realm of meanings, mind, logic," he writes; "the distributive occurs in the realm of existence, the discrete, the exclusive.... A good logic should use distinct symbols for different types of disjunction. It will then perhaps be more evident to extensional logicians that it is not possible to derive the distributive from the collective disjunction legitimately." The germ of truth in these contentions seems to be that the collective disjunction is to be handled metalinguistically, where we *mention* the disjunction *sign*, as in

> 'Tr ⌜(*a* Occ tomorrow v ~ *a* Occ tomorrow)⌝'.

In a metalanguage we speak of "meanings" and "logic" and perhaps of "minds." The distributive disjunction is appropriate at the object-linguistic level, the realm of "existence," of the discrete entities talked about. There is no need to distinguish the two disjunctions, however, as Weiss does, the distinction he has in mind being wholly captured at the metalinguistic level.

A better way of handling Aristotle's sea-fight tomorrow would be in terms of 'LTr' for L-truth. Where 'LogThm *a*' expresses that *a* is a logical theorem in the sense of first-order quantificational logic, we may let

> 'LTr *a*' be short for '(Tr *a* . LogThm *a*)'.

'Necessity' can then be handled by 'LTr', so that the form for the sea-fight sentence becomes

> '(E*e*)(E*a*)(SeaFight *e* . *a* Des *e* . LTr ⌜(*a* Occ tomorrow v ~ *a* Occ

tomorrow)⌝ . ~ (LTr ⌜*a* Occ tomorrow⌝ v LTr ⌜~ *a* Occ tomorrow⌝))'.

This is still too simple in some respects and can be improved upon.

Weiss's third thesis is that there is a distinction between an "extensional" conjunction and an "intensional" one. He comments that "nothing is so simple as 'and'. Yet it is a term used in multiple ways." Yes, indeed. Some of these have been studied in a previous paper.[8] Such study is still in its infancy. Here again, Weiss's distinction between two kinds of conjunction is adscititious. Many, many uses of 'and' are to be distinguished, but all the them may be accommodated within suitable metalanguages — even wholly extensional ones — in which intensions of various kinds may be handled.

Weiss speaks throughout of entities such as "propositions," as many another writer does and has done, but without giving an analysis of their internal structure.[9] We do not make much progress in logical and metalogical matters until we give up talk of propositions *sui generis* and in its place make fundamental use of the semantical truth-predicate and its environing theory.

Weiss's fourth thesis has to do with negation, that "a negation in thought is distinct from a rejection in fact." Of course it is, and there are many, many further types of negation also to be distinguished, as we have observed in I above. Weiss distinguishes only three. *Simple* negation for him seems to be essentially the usual one as symbolized by '~'. *Infectious* negation is that "in which the negating transforms the idea, meaning, or proposition to which it is applied (and therefore precludes the use of double negation as a device for getting back to the original)." This explanation seems in need of a considerable spelling out in an intensional metalogic. Surely, somewhere along the line, it can be handled in terms of the various devices discussed in I above. And, similarly, for Weiss's third kind of negation, *relational*, which is said to connect "opposing entities, $p \times q$, where q has the form of either an infectious or simple negation of p." This too needs explication in specific cases, and is probably best to be handled metalinguistically.

"The transition from one type of negation to another requires transformations which should be (but have not yet been) made the topic of extended logical inquiry." True, but beginnings have been made, and the basic notions needed for such study — including syntactical, semantical, and pragmatical ones — are now available. But of course we must distinguish, as Weiss does not, sentential negation, class- or virtual-class negation, thing-negation (as in mereology), event-negation (as in event- or progress-logic), relational or virtual relation negation, and so on and on. For each kind, special principles are available to characterize them. We should not lump them all together under some grand rubric and neglect consideration of the ontic specificities of the various kinds of entities needed.

In the fifth thesis, a distinction between "formal" and "material" implication is made, the distinction not being the usual one. "A formal implication is a structure connecting a given item in mind or language with some required other. The two items it connects are "at the same time," even when they are assigned different dates.... A material implication, in contrast, expresses how a potentiality in a thing will, by virtue of the structure of a thing and the world in which it exists, be actualized.... The distance between the two types of implication is traversed when we go from pure mathematics to physics — from a non-extended connection between meanings to an extended connection between times." These contentions are in need of a good deal of spelling out.

A quite new domain of applied logical theory emerges when times are explicitly taken to be values for variables, either primitively or by a suitable use of definitions. Formal implications, in Weiss's sense, can perhaps be regarded as being expressed with temporal quantifiers. If so, we should be told precisely how these quantifiers are to be allowed to occur and precisely what the "structure" of a formal implication is supposed to be. The only hint that is given is in the remark that "the astronomer now implies that the eclipse will take place a thousand years hence; the eclipse as taking place then is now understood by him the very way that he understands the premiss which implicates the consequence." The astronomer states *now* that on the given evidence expressed in such and such a premiss, such and such an event will occur a thousand years hence. (This statement is not, note, a formula of pure mathematics nor does it appear to be concerned with a "non-extended connection between meaning.") A "material" inplication, on the other hand, has something to do with potentiality and "extended connections between times." That the formulae of physics are of this character is far from clear.

Weiss is not one to show much respect for accepted terminology. Would it not have been better for him to have accepted the usual way of understanding formal and material implication, in terms of the *PM* '$\supset$', say, and the quantifiers? He could then have gone on to characterize his distinction by bringing in explicitly the additional notions needed — times, potentiality, actualization, structure, and so on. Each new item here is fraught with its own problems and these must be faced one by one. *Divide and conquer* should be a guiding methodological maxim here — as indeed *überhaupt* — as well as *Distinguer pour mieux unir*.

The next thesis, and indeed all the remaining ones, are concerned with acts of inference and thinking. "Thinking," we are told, "is an act of transforming an idea or judgement into another" and "is a specialized act of the mind which makes use of rules." The study of such acts is not strictly a part of logic, but rather of pragmatics, in which the human person and his relations to language are brought into consideration.[10] Let

'<Think> *e*' and '<Inf> *e*',

respectively, express that *e* is an act of *thinking* or of *inferring*. '<Think>' and '<Inf>' are so-called event-descriptive predicates or gerundives. Using these will help us to see how Weiss's further contentions may be developed and clarified.

The word 'idea' is so hoplelessly obscure that it seems better to avoid its use altogether. Hence, for the moment at least, let us think of thinking and inferring as having statements, or interpreted, declarative sentences, as their objects. "A soul inference is a thinking which terminates in the warranted qualification of an idea," Weiss tells us. Why not speak here, instead, of the objects or "items" qualified and insist that the inference *start* with a warranted "qualification" of such an object? More precisely, we could say that an inference is *sound* if it accords with the rules of the logic being employed and if its premisses are warranted as *true*.

Weiss is clear in his insistence, in the eighth thesis, that "inference is governed by rules, conformity to which determines the warrant of a transferred or transformed quality." "An inference may ... not only *transfer* [a truth-value from premises to conclusion] but also *transform* ... [it]; starting with *p* as true it can, to take the simplest case, end with not-*p* as false." There are also transformations involving modal values — or, as we would prefer to say here, involving LTr, LFls, and FTr and FFls (logical truth, logical falsehood, factual truth, and factual falsity).

After calling attention to these, Weiss skips on to some remarks about "the most productive modes of inference" being "found in the arts." It would be interesting to have been given some examples. It is difficult enough to characterize the modes of inference needed in the sciences, and also needed in discussion *about* the arts. But what rules of inference are to be found *in* the arts? The use of the term 'modes of inference' here seems out of place. Each art may well have "a form of warranted inference peculiar to it," but this is not one of inference in the usual sense. Better to use some new terms here appropriate to the arts in question, rather than to force a technical word upon all of the arts willy-nilly. (It is often said that the greatest defect of Nelson Goodman's theory of art is his unsophisticated, indeed, naive, use of 'denotation'.[11]) To paraphrase Weiss himself: "There is no point in trying to make one discipline conform to the rules [and terminology] of another. It is as foolish to demand that poetry conform to the canons of logic and to dismiss it as invalid or irrational because it cannot so conform as to demand that logic conform to the canons of poetry or to dismiss it as vapid or dead because of its failure to function as poetry does." Quite; let each have its own function, and let us develop a general vocabulary in which the various modes of functioning may be clearly described, individually as well as severally. At the same time, poetry has its hidden logic[12], just as logic has its hidden poetry — neither being revealed except to the highly initiate. As John Berryman remarked to the present author on one occasion: a good poem has as exact an intellectual structure as has a mathematical theorem.

Weiss remarks, in his ninth thesis, that "all thinking exemplifies rules and terminates in necessitated conclusions." And there are two kinds of rules, we are told: "contingent, alterable rules of transformation employed for some limited purpose; and necessary, unalterable rules.... A premiss and a contingent rule together suffice to give the conclusion. Given that 'Socrates is a man' and the contingent rule that 'all men are mortal', we can rigorously deduce ... 'Socrates is mortal'.... "This example shows that Weiss is not using 'rule' just in the sense of a rule of inference in the usual sense. 'All men are moral' is not a rule but a statement. *Valid* rules of inference are those studied in deductive logic. These are the "necessary, unalterable rules." A contingent rule of inference is of an altogether different ilk. Weiss comments that "the conclusions of the insane or of those who commit logical blunders are rigorously deducible by one who knows the premisses from which, and the manner in which, they think." A contingent rule is thus not strictly a rule, but is to be characterized in terms of a *pattern* of inferring which people might exhibit. (Recall III above.)

Weiss's remaining theses may all be viewed as having to do with patterns of inference and their accord or lack of accord with valid rules. To set the stage for the remaining theses, let us define '*p* Inf_e *a*,*b*' as '(<Inf>*e* . *p* Agent *e* . *e* From *b* . *e* To *a* . Sent *a* . Sent *b*)'.

The relation Agent here is that of being an agent of, To and From are the fundamental Russellian To- and From-relations previously discussed.[13] The definiendum expresses that *e* is an act of person *p*'s inferring sentence *a* from sentence *b*, where *b* is the premiss and *a* the conclusion.

An inference may or may not accord with those of logic. For specificity, let us take *modus ponens* (*MP*) and the rule of generalization (*Gen*) as the only logical rules — all others being derived ones, given the axioms of the system.[14] Let

'*a* Gen *b*' and '*a* MP *b*,*c*'

express that *a* results from *b* by one application of *Gen*, and that *a* results from *b* and *c* by one application of *MP*, respectively. The inferences of some people — their official "on the record" ones, anyhow — will be found to accord with these rules. Those of other people may not, but still may exhibit certain patterns or regularities, that might be of interest to the psycho- or socio-linguist studying actual usage rather than paradigmatic correctness. Even the insane, as Weiss notes, may have a "logic" all their own.

What does it mean to say that an act of inference *accords* with the *MP* and *Gen* of logic? Let us let

'*a* IC *b*,*c*' abbreviate '((*a* Gen *b* . *c* = *b*) v (*a* MP *b*,*c* . *c* = ⌜(*b* ⊃ *a*)⌝)'.

expressing that *a* is an *immediate consequence* of *b* and *c*. We may then let

'*p* $\mathrm{ValImInf}_e$ *a*,*b*' abbreviate '(*p* Inf_e *a*,*b* . (E*c*)(*a* IC *b*,*c* v a IC *c*,*b*))',

expressing that *e* is a *valid immediate inference* of *a* from *b* by person *p*. The definition of the notion of being a valid inference in general, as a sucession of steps of valid immediate inference, requires some additional technicalities that need not be given here.[15]

Weiss's remaining theses, the tenth through the eighteenth, all seem to rest upon the distinction just drawn, between steps of valid inference in accord, say, with standard quantificational logic, and steps of inference that are not valid in this way — allowing however, that they might exhibit certain idiosyncratic patterns of their own and thus be "valid" in some Pickwickian sense. Thus Weiss contends that "a particular conclusion has content, meaning, implications beyond that expressed in the general rule." This seems to mean that "the process through which one goes in order to get a conclusion, connected with a premiss by a necessary rule of logic, has content, power, a meaning not altogether expressed in such a rule." Quite. The "content, power, meaning" of the conclusion is of especial interest, no doubt, to the agent of the inferring; his acts of inferring aim toward it, so to speak; to obtain it is the purport of his performing the acts. And what he arrives at may, of course, in given cases, surprise him, so that "a process of moving to a conclusion ends [or may end] with what is not wholly predictable," and thus "all inference is logically necessitated [in one sense or another] and creatively free."

Clearly "[valid] inferences and productive processes [of moving to a conclusion] are distinct," but the latter may accord with the former on suitable occasions. Further, "inferences make a difference to what is implied," in the sense that "inference is an art in which one risks replacing a satisfactory premiss by an unsatisfactory conclusion." Also, "processes [of carrying out the steps of an inference] add to what is materially implied," in accord with the Pickwickian "rule" of inference governing it.

Inferring, whether valid or not, takes time. The successive steps succeed each other in such a way that the earlier ones bear the temporal relation of occurring *before* the later ones. Also there is a kind of "parallel between formal implications with their corresponding inferences and material implications with their corresponding processes...." It is also noted that "there are invalid inferences, but not invalid processes." Processes take place just as they do, and, as already observed, "formal implications and processes are not necessarily in accord."

All of these points — pragmatical ones, note — may be expressed in terms of 'Inf', 'MP', 'Gen', and so on, together with such additional expressions as might be needed for stating any Pickwickian rules under consideration.

Let us glance briefly now at the salient points of Weiss's earlier papers in logic, those from 1928 to 1933, one by one.

In "The Theory of Types"[16] Weiss reiterates the well-known point that, on the basis of the Whitehead-Russell theory, "it is impossible or meaningless to state propositions which have an unrestricted possible range of values, or which, in any sense, are arguments to themselves." To bring Weiss's discussion up to date, we should have to avoid all reference to such hoary objects as "propositions." Such objects should be forbidden *ab initio* in all logical and philosophical discussion, but perhaps introduced by definition step-wise.[17] At best, propositions correspond with, or are expressed by, declarative sentences in some fashion. The analysis of their structure must thus presuppose a good deal of semantical (or pragmatical) theory in need of development. It will not do, therefore, to proceed as though we had any clear notion of proposition to begin with, as Weiss, following Russell, Leśniewski, and a host of others, has done. If we face up to semantical matters, we will find that propositions at best are very complex constructs, so complex in fact that the very purpose of introducing them at all seems best served by simpler means.

Weiss does not distinguish clearly what has come to be known, following Ramsey and Chwistek, as the distinction between the simple theory of types and the ramified theory. The latter is needed for handling the so-called semantical paradoxes — that of Epimenides, heterologicality, and so on — from the purely logical ones — concerning *inter alia* the class of all classes that are not members of themselves. This distinction has become more or less standard and Weiss's paper seems to point toward it. He purports to draw a distinction, moreover, between "restricted" and "unrestricted" propositions in order to gain a kind of "restricted" theory of types. Even so, one would surely need to distinguish object-linguistic from metalinguistic matters, and to treat the two kinds of paradoxes quite differently.

Thus is not the occasion to examine Weiss's resticted theory in its own terms. Much semantics would have to be accommodated in those terms, and many distinctions made, but all this would have to be carried out by someone friendly to self-reference. The view here is that the possibility of self-reference in natural language should not be taken as a guiding principle for logical investigation.[18] At best, self-reference in natural language is severely restricted, and can perhaps be handled as extended usage in some suitable fashion. Methodologically, according to the view here, non-self-referential logic comes first and the study of grammar must bow to its authority; and we should bear in mind that the scope of that authority is, almost daily, being extended in the exact study of logical form, in which self-reference plays no essential role.

In "Relativity in Logic"[19] it is urged that there are many logics, and types of logic, to be distinguished, each in some sense as good as any other. It is contended that "a logical system need not have an application and that non-applicable systems are as much the logician's study as the different possible geometries are the mathematician's, ... [and] the different systems of logic should be differentiated on some other ground than their possible application to our usual 'legitimate' trains of reasoning. It should be indifferent to a logician whether his system can be used." As opposed to this view, one can always insist upon some "interpretation," however bizarre or *outré*, some "model" however queer. All of these interpretations, however, are intended by the logician or mathematician to serve some significant intellectual purpose or other, to help solve a practical problem, or to help somehow in the enterprise of knowledge.

Weiss invites us to consider "values" other than truth and falsity. Necessity, possibility, and contingency are hinted at, but a third value, intermediary somehow between truth and falsity in the sense of Łukasiewicz, is not. Nor any fourth value in the sense of four-valued logic; and so on. Weiss considers several ways of construing 'implication' truth-functionally, other than in the sense of "material" implication. But since he does not care whether any of these are "applicable" or not, no constraints are at hand in the light of which to approve or not. Further, all of these types of "implication" may be introduced within the *PM*, two-valued framework, so that nothing essentially new emerges from the consideration of them. They all reduce to the same old thing, looked at in a slightly different way.

Fruitful research in logic is always governed by a certain sparseness: clearcut aims, sharp techniques, ontologies cut to the bone. Without such sparseness, there will be a too-muchness, aims that are not clear, ill-worked-out techniques, ontologies that are too rich for the purposes at hand. With a too-muchness, one may fail to see problems that cry out for consideration. In this paper, Weiss is willing to speak of "entities" and notions of various kinds, without an analysis of them and without showing how they are interrelated. A few such are: propositions, entailment, truth, falsity, internal and external relations, assertion, contingency, necessity, necessary factual correlation, validity, property-value, categorical value, and so on. Weiss uses some of these terms in his own way. Some of them he generalizes to the utmost; but to what avail? He ends up with a notion of *The Logic* (with capitals) "which has an infinite [denumerable or non-denumerable (?)] number of elements capable of an infinite number of property-values. It can never be expressed." Would it not be more helpful here to have put forward a sparser logic, of some significant applicability, capable of being expressed? Such a logic was given — rather well, be it said — in the monumental *PM*, which Weiss seems to reject much too peremptorily.

In "The Nature of Systems"[20] many valuable points are made, but, again, without bringing in explicitly metalogical considerations to provide the wherewithal for doing so adequately. Historically speaking, this paper may be viewed as an attempt to develop a theory of systems, to "characterize all possible systems, to justify their methodology," and thus to provide some headway "in an important though neglected [area of] study." Straightaway we are told that "a system is a set of propositions, some of which (the postulates and primitive propositions) are derived by means of substitutional and inferential rules." Here again we must be wary of proposition-talk in the absence of a metalinguistic characterization as to how they are expressed in sentences. Also the use of 'set' here is unwise — although widespread — in view of the kinds of considerations addressed in V-X above.[21] Further, Weiss is as much interested in uninterpreted systems as interpreted ones. As a matter of fact, historical and otherwise, uninterpreted systems are not of much interest in either logic, mathematics, the sciences, or philosophy, as has already been noted above. Only so-called *semantical* systems are of such interest, those with explicit *semantical rules* giving some significant or interesting interpretation. Of course, a *syntactical* system may be at hand prior to the formulation of such rules, but then the system will have been formulated with a view to adding such rules. Even after such rules have been added, the resulting system might be "abstract" and descriptive of mere possibilities — but with the semantical rules we will be told at least what these possibilities are.

In "Two-Valued Logic — Another Approach"[22], Weiss makes some trenchant commments purported to bolster his general conception of The Logic (with capitals) as a subject-matter rather than as an instrument. That logic should be one or the other — or both, as with Avicenna, for example — has been a running debate for many centuries. "It is one of the paradoxes of modern logic that most symbolic logicians are not really interested in logic at all." (How can they be, in Weiss's sense, unless there is a special subject-matter for them to be interested in? It behooves Weiss to convince us that there is one.) "Their concern is with mathematics, epistemology, system, symbolism, language, etc.... They pay practically no attention to the question as to the nature of logical propositions. One would have difficulty in discovering from their writings what it means to say that logical propositions are necessary true." What are "propositions" anyhow, and what then in particular is a "logical proposition"? The urgency of the former question has been indicated above, and the latter surely cannot be answered except on the basis of some answer to the former. And, of course, logicians are interested in "mathematics, epistemology," and so on, as they should be — in fact logic takes all knowledge for its province, just as metaphysics does. We should not condemn logicians for the generality of the applicability of their subject, and withhold condemnation of metaphysicians for the generality of theirs. At the same time, there may well be a sense in which metaphysics is a more general pursuit than logic. Perhaps The Logic is being forced here to take on the tasks of the latter.

Weiss writes about 'necessarily true' as though he knows what it means. This was done prior to Quine's important paper "Three Grades of Modal Involvement" and to the present author's "Does Modal Logic Rest upon a Mistake?", which bring to the fore the difficulties connected with attempts to characterize that notion other than in clear-cut semantical terms.[23] Weiss writes in the mode of "inarticulate certainty" on semantical matters, where we have, fifty-odd years later, only "articulate hesitency."

Weiss contends that "all logical propositions are on the same level and are of equal generality." If this is supposed to say that all logical truths, in the sense of 'LTr', say, are of the same generality, it is not the case. Clearly

$$\text{`(B Cora}_P \text{ v} \sim \text{B Cora}_P\text{)'},$$

where 'Cora_P' stands, again, for Peirce's horse, and 'B' for a shade of brown appropriate for horses, is not of the same generality as

$$\text{`}(x)(\text{B } x \text{ v} \sim \text{B } x)\text{'},$$

a more general form, for a law of excluded middle concerning 'B'. But the two are on the "same level" in the sense of both being LTr.

Also, as a matter of historical fact, logicians have been paying considerable attention to the problem of how validity, logical truth, necessity, and the like, are best to be characterized, in the years since Weiss wrote. But much of it has been done at the metalogical level. Weiss's paper has the merit of calling attention to such topics, which, however, can be adequately approached only with further notions or tools, so to speak, that were just coming on the scene in Europe at the time he wrote.

In "The Metaphysics and Logic of Classes"[24], Weiss puts forward four more objections to *PM*, as follows. The first is that the notion of class "is actually, though not symbolically, defined by using the notion of a class in the definition." Apparently Weiss fails to take into account the definition *20.01 in which symbols for classes are defined contextually. No symbol for class, however, occurs in the definiens. (The '$\phi!\hat{z}$' that occurs there is used, as Russell explains, "when we wish to speak of the function itself as opposed to its arguments.") To call attention to this is neither to presuppose nor defend Russell's explanations, which are no doubt confused; it is merely to point out that Weiss's first criticism here seems not to be fully justified.

The second objection is that "the notion of a class is reached through an analysis of the propositional function" and that "because he was dominated by a metaphysical view that individuals are independent of and more basic than the universals supposed to qualify them ... Russell dealt only with those propositional functions which had variable subjects." (Strictly there are no such things as "variable subjects"; there are only subject variables. The very phrase 'variable subject' rests upon a confusion of use and mention.) But in 'ϕx' or '$\phi ! x$' the 'ϕ' is as much a variable as the 'x' is, but is of one type higher than whatever the type of 'x' is. Weiss seems to forget here the device of "typical ambiguity" that is used throughout *PM*. Also, surely it is not the case that Russell — at least in *PM* — regarded individuals as either "more basic than" or as "independent of" the "propositional function" they may or may not satisfy. Both kinds of entities have equal status as values for the variables of the system, albeit these variables are of differing types.

The fourth objection also seems to rest upon a confusion of use and mention. "A class is held to be an incomplete symbol or a fiction, and a member of a class is related to its class by ε. But if the members of classes are real, in what sense can they be related to incomplete symbols or to fictions? And if numbers are classes of classes similar to a given class, what are the objects of mathematics but incomplete symbols of incomplete symbols, or fictions of fictions?" The answer is that a "class" is *not* an "incomplete symbol." Russell is explicit that "the *symbols* [italics added] for classes, like those for descriptions, are in our system, incomplete symbols, ..., [and] classes, so far as we introduce them, are merely symbolic or linguistic conveniences, not genuine objects as their members are if they are individuals." Classes are thus "fictions" for Russell only in the very special and harmless sense of being introduced by contextual definition.

Weiss's third objection might be thought to fare rather better. It concerns the duplication of the cardinal numbers at each type level $\geqq 3$, where individuals are taken as of type 1. This duplication has often been remarked upon and thought objectionable. Weiss's description of his objection, however, is not quite accurate. According to him, in *PM* "any number can be said to be a member of members of any other number, as well as of itself." This latter is not quite correct — no number can ever be said correctly to be a member of itself, according to the theory of types. The former contention is correct only if we distinguish carefully the type of a given cardinal number. Thus 2_3 is the number 2 of type 3, 2_4 of type 4, and so on. Let $\{\alpha_n, \beta_n\}_{(n+1)}$ be the class of type $n+1$, its only members being α_n and β_n each of type n. Then it is correct to observe, as Weiss does, that

$$(E\alpha_4)(2_3 \ \varepsilon \ \alpha_4 \ . \ \alpha_4 \ \varepsilon \ 1_5) \ . \ (E\alpha_4)(2_3 \ \varepsilon \ \alpha_4 \ . \ \alpha_4 \ \varepsilon \ 2_5) \ . \ \ldots \ .$$

and similarly for 3_3, 4_3, and so on. And similarly for cardinals of higher type. But note that this adds up to saying only that cardinals of type n are members of members of cardinals of type $n + 2$.

Weiss thinks that all this is "at best a monstrous doctrine." But why? It is rather a fact of life for the simplified theory of types, and one inherited by all modern set-theories in which type-theory may be imbedded.[25] Further, an allied kind of duplication is a fact of life for other kinds of adjectivals in natural language. It seems generally accepted now that 'large', say, is to be parsed in terms of 'large for elephants', 'large for jets', 'large for porpoises', and so on. The admission of such a duplication is almost a necessity for gaining clear-cut logical forms for the appropriate natural sentences. And similarly, then, for natural sentences containing numerical modifiers. Even so, no account of these latter would be adequate without a theory of acts of counting and of measuring trailing along.[26]

In the bulk of this paper Weiss puts forward a theory of his own, as based on unit classes, classes with one member, as fundamental, and "all classes having more than one member are defined [in terms of expressions for] ... sums of unit classes." A very considerable mathematical theory about such sums is then needed, in order to provide for denumerable sums, as well as all manner of non-denumerable ones. And, similarly, all manner of relations must be provided for, about which Weiss tells us singularly little. The technical details here loom large, and perhaps would be similar in some respect to those of V above. Weiss does not develop his view even to the point of providing a notation in which a single mathematical principle could be proved or a single definition given. Even so, the hints he gives might well be worth developing in terms of an exact formulation, even now, fifty-two years after it was written — but without presupposing the notion of "proposition" unless given a full and adequate semantical analysis somewhere along the line.

In "The Metaphysical and the Logical Individual" of 1933[27], a distinction is drawn between a "logical" individual, taken as "an argument to a propositional function, such as "... is red" [and] a metaphysical individual ... [as] the indivisible, existential unity from which both logical individuals and their characterizing predicates are derived by abstraction." The distinction, essentially that of the 1957 paper, seems adscititious. Why cannot a "metaphysical" individual itself be taken as "an argument to a propositional function" — provided we have a clear notion of this latter? For Weiss, "the logical individual is simple, unanalyzable, and non-existent," whereas a "metaphysical individual is a what imbued with existence, a that suffused with intelligibility, characters before they are shorn of their existence; existence before it has been sundered from that which exists; the source of the complex and of the simple." The "logical" individual is thus a sheer fiction, of no use anywhere, neither within nor outside of a logical system. Logic is not so much a subject-matter as a framework. *Logica utens*, logic in the context of an applied system, as we have already in effect observed, is the only kind worthy of serious pursuit. Weiss's notion of a "logical" individual, it is to be feared, like a "proposition," and the like, must inhabit a Disneyland of unreality far removed from, and of no help to, the common toil of trying to get our ideas clear, of removing the debris that stands in the way of gaining knowledge, and of keeping to the much-prized Einsteinian "logical simplicity that helps to keep us on the right track."

In "On Alternative Logics"[28] Weiss is concerned mostly with some comments by C.I. Lewis[29] concerning implication and the multiplicity of alternative logics. Because this whole subject has become wholly swallowed up in the subsequent development of metalogic, semantics, pragmatics, and the like, the details of the discussion lie beyond present concern. It will be more interesting to comment on the last paragraph of the paper, which is in effect a plea for a single logic as somehow basic, an echo of Whitehead's plea for "one God, one country, one logic."

Weiss points out that "there is no need for believers in multiple logics to invent interesting diagrams [or matrices] or novel symbols to demonstrate their point.... All they need to do is to analyse a *single, valid* logical proposition and show that when they take it extensionally, ..., or that when taken intensionally, the meaning of the antecedent does not contain the meaning of the consequent. Until this is done the whole case for the pragmatic choice of logics, for the untrammelled freedom of the mind, for the infinite logics to be uncovered by the man who can free himself from worldly worries, will remain a position stated and not a doctrine proved."

What, now, does Weiss understand here by taking a logic extensionally as over and against taking it intensionally? The difference is perhaps related with his distinction between "formal" and "material" implication, which we have already met above. In any case, here we are told that "the error in believing that there are many logics because there are many truth-values [necessity, contingency, and so on, in addition to truth and falsity], has its roots in the confusion between an extensional and an intensional approach to logic. When we take the extensional approach we are interested in knowing what can be necessarily concluded from a knowledge of the truth-value of a proposition or set of propositions.... From 'this is red' an extensional logician would conclude 'this is red or this is not red' [or even 'this is red or this is blue'?].... He cannot conclude 'this is a color' from 'this is red' unless 'this is red' is understood as 'this is red or green or blue, etc.' ... From the intensional side, our interest in logic is confined to knowing what propositions or concepts have their meanings contained in a given proposition or concept.... From 'this is red' we conclude 'this is a color', 'this is darker than pink', 'this is extended', etc...."

The key item to observe here is that we can handle Weiss's "intensional" approach within an extensional system, by adding to the system *meaning-postulates* in Carnap's sense.[30] Let us first note that 'this is red' is ambiguous, depending upon whether the 'is' is taken as the 'is' of predication or the 'is' of identity. Further, the presence of 'this' brings in all the difficulties concerned with ostension. To simplify, let us construe the 'is' in the sense of predication, and think of 'this' as indicating some specific present object before us. We could then have meaning-postulates added to the system that

$(x)(x \text{ is red} \supset x \text{ is colored})$,
$(x)(x \text{ is red} \supset x \text{ is extended})$,
$(x)(x \text{ is red} \supset x \text{ is darker than any pink object})$,

and so on. The logic of 'this' would then be presumed to justify an assertion of

'(this is red ⊃ this is colored)',
'(this is red ⊃ this is extended)',

and so on, from the assumption of these meaning-postulates. In this way Weiss's "intensional" forms of inference could easily be handled.

By the introduction of further devices, many of them metalinguistic, all manner of other locutions, commonly called 'intensional' may be handled, giving support to the supposition of there being only one basic, two-valued, logic of which all "alternatives" are specializations in some fashion or other.[31] The gist of Weiss's discussion here seems to point to this supposition. This one logic, however, is still to be distinguished from The Logic as being formulable or expressible and in not usurping the tasks of a metaphysics.

Weiss's contribution to the Seventh International Congress of Philosophy, at Oxford, in 1935, was entitled "Entailment and the Future of Logic."[32] It has been saved for the end here because it does to some extent summarize most of his work in logic. And recent developments have in fact taken place in just the areas to which Weiss called attention.

The discussion of "entailment" and "implication" here, as in the other papers, may be carried out in terms of the standard, two-valued framework. Further, Weiss does not develop his ideas far enough to include explicit consideration of the quantifiers. Scarcely is a quantifier mentioned or even alluded to. He seems to think of logic as having to do only with the connectives 'and', 'or', 'not', and so on. This is of course a very serious limitation. As a result, the strictures of *PM* seem unjustified here, as in Weiss's other papers.

Here again attention is called to intensional logic as having to do with *meanings* and "with expressing the relationship of *meanings* to one another." The notion 'meaning' is notoriously slippery and must be handled with great delicacy and care as a technical notion. Weiss comments that it would be desirable in the future of logic, to develop "a complete intensional logic ... [that] would allow for the assertion of propositions not now [in 1930] acknowledged, and for a possible derivation of and a detailed comparison with extensional logics." Yes, indeed, and much work has been done in this direction — whether successful or not being a matter of opinion — by Carnap and Alonzo Church and others, mainly in the wake of Frege's seminal "Über Sinn und Bedeutung" of 1892. It is a pity that Weiss was not familiar with the contents of this latter paper.

Weiss's work in logic must be viewed in the light of the history of work done in this country just prior to the arrival on the scene of the metalogical contributions of Leśniewski and Tarski and of the work of Carnap and of the Hilbert school. As concerns philosophic logic — to be contrasted sharply nowadays with mathematical logic — the development of logical syntax and semantics was of high relevance, to which was added an approach to the systematic pragmatics and to process- or event-logic some years later. Weiss's work came along just a few years too early, before the newer Continental work had been absorbed. It inevitably must suffer in comparison. Even so, it contains many highly perceptive *aperçus*, the full character of which can best be assessed and seen in the light of these newer developments. In short, Weiss must be credited with having raised many interesting and important philosophical issues connected with logic at a time when the techniques appropriate for handling them were not available.

CHAPTER SEVENTEEN

On Quine's "facts of the matter"

"On spécule sur tout, même sur la famine."

Never has a philosophical paper started out more felicitously than Quine's "Facts of the Matter."[1] At the beginning, the reader is tempted to agree with every point. The prose is perhaps the finest in philosophical writing in English since that of Gilbert Ryle. Quine recalls at the outset Horne Tooke's *The Diversions of Purley* of 1786, in which it was urged that Locke's *Essay* would have fared better if only the word 'word' had been substituted throughout for 'idea'. Indeed, yes. The linguistic turn in philosophy might then have taken place more than 250 years earlier than it actually did. Of course a good deal of vital material in Locke's work would have been lost, but there would also have been much gain. "We think mostly in words," Quine notes, "and we report our thought wholly in words." Well, not quite — there are grimaces and gestures to be accounted for, works of art and other non-verbal media of communication, and so on. Even so, Quine is right, surely, in urging that we should "seek a firm footing ... in words" and thus should "take one leaf from the old-time [Lockean] philosophy and another from John Horne Tooke. Philosophical inquiry should begin with the clear, yes; but with clear words."

How can we "begin" with clear words if in fact they are not clear to begin with? Quine outlines his well-known views about "the social character of language and the use of language in communication," about how we come to accept occasion-sentences (or to reject them, or to abstain from either accepting or rejecting them), about subjective similarity or "similarity by the subject's own lights," about how we build up an observational vocabulary, and so on. This is all set forward with admirable clarity. "And now what happens," Quine asks, "when at last we can be said to use some of these [observation] sentences or parts [segments?] of them as terms denoting some sort of supposed objects? The main thing to settle ... is their individuation; we have to fix standards of sameness and difference. [And] ... at this point [in our learning] little or no attention will be paid differences of [the perspectives] ... from which they are observed.... What are posited as objects for the terms to refer to will be, primarily, objects that are accounted identical under changes of perspective. This explains the primacy of bodies.... If inquiry is to begin with what is clear, then let us begin as physicalists."

There is quite a leap in all this, it should be observed, from the learing of the observational vocabulary to the "positing" of objects. And quite another leap from positing objects to anything sophisticated enough to be labelled 'physicalism'. The first is akin to the leap from Berkeleyan "ideas" to the reism, say, of Kotarbinski, and the second, from this latter to a full-fledged Carnapian physicalism. Neither leap is undertaken by the learner as such. Quine is forcing them upon cognition-theory, or the theory of learning, without ever so much as even mentioning or taking account of important recent experimental and conceptual work in the field. The second leap, from reism to physicalism, he is more persuasive about, and we shall come to it in a moment.

Along the way of mastering language we encounter general terms. "Grammatical analogy between general terms and singular terms encourages us to treat a general term as if it designated a single object, and thus we come to posit a realm of objects for the general terms to designate: a realm of properties, or sets. What with the nominalizing also of verbs and clauses, a vaguely varied and very untidy ontology grows up." Indeed it does, and Quine, alas, seems not to face up to its full complexity. The variation may not be so vague after all, and we should work to try to make the ontology more tidy. The language of science is an "object-oriented idiom," Quine contends, but he has in mind the stale science of yesteryear. Almost daily the languages of the sciences are becoming less and less "object-oriented," as recent work in physics, chemistry, biology, neurology, and psychology are showing.[2] In any case, the leap from Quine's "objects" to the positing of a realm of *sets* is a third leap, over a chasm that requires an enormous effort to bridge, both from the point of view of the learner and from that of the theorist of learning.

Another stupendous leap in Quine's thought is from the notion of set to the "regimentation" of natural language. His views on this matter are familiar. "Language ... [as] regimented has a simple grammar [sic!]. There is a lexicon of predicates. Each atomic sentence of the language consists of a predicate, say, an n-place predicate, adjoined to n variables. The rest of the sentences are built up of the atomic ones by truth functions and quantification." But all this is much too simple. For no domain of science whatsoever, is an adequate "regimentation" of the kind Quine speaks — where its variables are taken to range over "objects" — forthcoming without a considerable ado, if at all. It is not clear that a fully adequate formalization for even classical particle-mechanics has been given, and surely none is forthcoming on the basis of present knowledge for quantum mechanics. Quine leaves this latter aside "in order not to complicate the picture." For similar reasons he must leave aside most portions of other areas of physics, and of the other sciences as well. And that natural language as regimented has a "simple" grammar is surely news to anyone who has worked on the subject at any depth.

Quine has always seemed to be somewhat wary of positing sets. "Much of the positing of abstract objects [such as sets] that seems to go on in ordinary language proves to be gratuitious and eliminable, but much of it also proves valuable." He comments, further, that "how *sets* can pay their way is classically illustrated by the definition of the *closed iterate* of a two-place predicate. Ancestor, for instance, is the closed iterate of parent." No, it is rather a two-place *predicate*, 'ancestor of', that is the closed iterate of the two-place *predicate* 'parent of'. And, of course, sets can pay their way also by providing for mathematics, and this would seem to be the most defensible reason for positing them.

Quine goes on to a paragraph that — really for him — is not clear. "It must be emphasized," he writes, "that when we reckon the ontology as comprising just the values of the variables, we are assuming the strictly regimented notation: just predicates, variables, quantifiers, and truth function. Admission of additional linguistic elements can upset this ontological standard. Thus suppose someone adopts outright an operator for forming the closed iterates of predicates, instead of defining it with the help of an ontology of sets. Are we to say that he has saved on ontology? I say rather that he has shelved the ontological question by switching to a language that is not explicit on ontology. His ontology is indeterminate, except relative to some agreed translation of his notation into our regimented one."

The contentions here are curious. The person who adopts as a primitive a notion "for forming the closed iterates of two-place predicates, instead of defining it with the help of an ontology of sets" will surely have "saved on ontology" provided he formalizes his theory without sets at all as values for variables.[3] Here there is real gain in ontological economy — and he will of course give axioms or axiom-schemata governing that new primitive. But if he introduces an expression for designating the operator as a primitive, in a theory in which sets are also available, he will not have changed the ontology at all. The situation is similar to taking '=' for identity as a primitive. If sets are available, there is no need to do this; if not there is.[4] In either case, one has not "switched to a language that is not explicit on ontology." But if one has taken '=' as a primitive in a (first-order) theory without sets, it is surely not correct to contend that one's ontology is thereby rendered "indeterminate." Is Quine stating here that one would be committed to sets in any such theory? He would seem to be by the added clause that the ontology is then indeterminate "except relative to some agreed translation of his notation into our regimented one." This last contention seems rather extreme, as though no alternative agreed-upon translation into any other theory or kind of theory is ever to be condoned.

"Another way in which quantification over sets or numbers or other abstract objects can sometimes be avoided is by admitting a modal operator of necessity," we are told — with a footnote reference to Hilary Putnam's "Mathematics without Foundations"[5] — "if we can see our way of making appropriate sense of the device. Here again we are presented not with an ontological saving, but with a question of foreign exchange." Quine seems to condone here Putnam's method for avoiding reference to abstract objects if sense can be made of the notion of necessity. That Putnam's method is unsound and highly dubious has been shown in detail elsewhere.[6] Here, indeed, we do seem to be presented with fraudulent currency, and it behooves Putnam to make good the exchange.

If, as Quine tells us, "for some sentences there is no acceptable regimentation at all," is not the programme of regimentation doomed right at the start? Further, so we are told, "regiment as he will, he [the translator] cannot make do with just bodies;" he must admit sets also. This pronouncement seems also to be rather *ex cathedra*, and runs counter to the contentions of V above. We shall reflect upon these later, in connection with the leap to physicalism.

What matters for Quine's physicalism, is that "where positions and states of bodies do not matter, there is no fact of the matter." But even positions are states, states of being in such and such a position. Quine suggests that he should perhaps "better phrase the matter in terms of *change*: there is not change without a change in the positions or states of bodies." How are these best to be handled? In "latter-day" physics, Quine points out, "there are indications that the utility of the particle model, the extrapolation of the primordial body into the very small, is now marginal at best. A field theory may be more to the point: a theory in which various states are directly ascribed in varying degrees to various regions of space-time. Thus at last bodies themselves go by the board — bodies that were the primordial posits, the paradigmatic objects most clearly and perspicuously beheld."

Quine's "brave new ontology" is thus to consist of "real numbers, needed to measure the intensity of the various states, and ... [of] the space-time regions to which the states are ascribed. By identifying each space-time point with a quadruple of real or complex numbers according to an arbitrary system of coordinates, we can explain the space-time regions as sets of quadruples of numbers. The numbers themselves can be constructed within set theory in known ways, and indeed in pure set theory; that is, set theory with no individuals as ground elements, set theory devoid of concrete objects. The brave new ontology is, in short, the purely abstract ontology of pure set theory, pure mathematics." All of this, of course, harks back, in essentials, to Carnap's discussions of coordinate languages.[7] Let us accept it *pro tem* in its own terms, and abstain from questioning whether this kind of view can be harmonized with "latter-day" physics. It is to be feared that it cannot, but the point need not be pressed here.

Quine thinks that "a lesson to be drawn from this debacle is that ontology is not what mainly matters.... Sentences, in their truth or falsity, are what run deep; ontology is by the way. The point gains [— or does it rather lose? —] in vividness when we reflect on the multiplicity of possible interpretations of any consistent formal system.... The sentences of ... [our coordinate] language that are true under countless reinterpretations of the range of values of the variables.... Perhaps then our primary concern belongs with the truth of sentences *and with their truth conditions* [italics added], rather than with the reference of terms. If we adopt this attitude, questions of reference and ontology become quite incidental. Ontological stipulations can play a role in the truth conditions of theoretical sentences, but a role that could be played as well by any number of alternative ontological stipulations...."

It is not that ontological stipulations "can" play a role in the truth-conditions of a sentence, it is that they *must* if 'true' and 'false' are given any *adequate* definition in the semantics of the language. Of course, alternative ontologies will do as well, but *quā* physicists and philosophers of physics, it is some one basic, physical interpretation that is of interest to us. The ontology of this interpretation is by no means incidental, but in fact as fundamental to the theory as the inner mathematics of the theory itself. Sentences as true or false "run deep," of course; but in saying this we must understand 'true' and 'false' in the only "deep" sense of relevance, that of theoretical semantics. Moreover, the sentential forms of the language are "true of" or "false of" such and such objects. Quine is not proposing that we accept the system merely as a *syntactical* one; no, he surely intends to use 'true' and 'false' as *semantical* terms, so that the system envisaged is intended to be a semantical one.

Bridgman was tireless, years back, in pointing out that "for a complete appraisal [and understanding] of what is contained in any [physico-] mathematical theory an analysis of the text is just as essential as an analysis of the equations. In particular, the text contains the unanalyzables of the theory and thus involves its essential limitations" — including the limitation to a given ontology. The text gives "what the significance of the equations is." As Bridgman noted, "a complete mathematical formulation involves equations plus text.... The necessity of a text is almost always overlooked, but I think it must be recognized to be essential, and [italics added] *a study of what it must contain is as necessary for an adequate conception of the nature of the [physico-] mathematical theory as a study of the equations themselves.*"[8] The modern way of providing for the "text" is within an explicative semantics, the explication giving the intended interpretation of terms of the intended ontology.[9]

Let us recall the very statement of the Adequacy Condition — Tarski's paradigm (T) — concerning the semantical truth-concept as defined for a given system. This is, it will be recalled, that every sentence of the metalanguage of the form

(T) $\ulcorner \mathrm{Tr}\ \boldsymbol{a} \equiv \text{-----} \urcorner$,

be provable (or true or at least "holds" in metalanguage), where '-----' is a sentence of the language and ***a*** is a shape- or structural-description of it or of a metalinguistic translation of it. Where (T) holds, we say that '-----' is a truth-condition for the sentence ***a***. If the semantics is an explicative one, for Quine's kind of system, '-----' will be equivalent to some statement of the system about points and space-time regions, even though these are not themselves values for the variables of the system itself. Informally, Quine "identifies" points with quadruples of real or complex numbers, but in the explicative semantics, the "text," this identification must be made fully explicit. The physics is not contained just in the equations but in the equations-cum-text. The conditions for the truth of the equations (or sentences within the "pure" set-theory with its roster of physical predicates) will have to contain variables over space-time points and regions. Without this, Quine's "physicalism," that "there is no difference in matters of fact without a difference in the fulfillment of the physical-state predicates by space-time regions," could not even be stated, and would indeed have to go by the way.

In this present paper, it should be remarked again, we have neither accepted nor rejected Quine's theory, but merely tried to make sense of it in semantical terms. Let us go on now to show how the essentials of his set-theoretic physicalism — let us dub it such — can be more appropriately accommodated within a given interpretation of the material of V above.

Let us think of the "individuals" of the homology of V as being space-time regions in Quine's sense. Of course 'space-time', 'space', and 'time' are not taken as technical terms — they become so as constructs only in the peroration of the theory. The part-whole relation is then the part-whole or inclusion relation between regions. The "cardinal units" of the theory become unit regions, not further divisible into smaller ones, and are such that physical magnitudes may be ascribed to them. In this respect they are unlike geometrical points, to which such magnitudes are not ascribable. In terms of the ordinal individuals and Boolean sums of them, attributive relations for physical states as applied to regions may be introduced, as well as non-attributive relations for relations between or among states. The various kinds of number may then be provided for essentially as in "On Mereology and the Heroic Course"[10], quadruples of real numbers then being available to play the role needed. Whatever expressions for primitive predicates are needed in Quine's treatment may be introduced now as primitive names of relation-individuals. In this way the full expressive power of his system is achieved within a unified framework. Physics and mathematics are brought together. No grand leap is needed from one to the other, no unbridgeable saltation from space-time regions to some mysterious new kind of entity called 'sets'. For the first time ever, perhaps, physics and mathematics can live together harmoniously under the same roof without the assumption of a brave new — but foreign — ontology of sets.

Quine's envisagement is, of course, mere programme. All kinds of difficulties are to be expected when one gets down to the real nitty-gritty of the details. One might think that this will all be a mere "routine" matter of smooth sailing. But no. There are the separate areas of physics itself that cannot be easily unified on the basis of present knowledge. And there are internal difficulties in each that must be faced up to squarely when an even partial formalization is attempted. Philosophers may step in where working scientists fear to tread — of course — but by carrying on where they leave off, in showing in detail how deeply philosophical notions permeate the languages of science and how awareness of logical matters helps in gaining better formulations and greater clarification and hence understanding — in short, by cooperating in the enterprise of science *from close to* rather than from afar off.

Suppose Quine's programme could be carried out successfully, at least to some extent. It might appear that this would be to the total neglect of so-called "mental" phenomena. Among the state-predicates, however, he purports to include mental ones. "We have two species of predicates, mental and physical, but both sorts apply to bodies," or, in the "pure" set-theoretic rendering, presumably to space-time regions. Physics, in its present state of development, apparently has no need of mental state-predicates, although neurophysiology apparently does, as the work of Sir John Eccles and R.W. Sperry *inter alia* has apparently shown. The day may well come when the use of such predicates within the language of science will have become a *fait accompli*. Metaphysical idealists have devoutly been wishing for such a consumation all along.

Quine makes a good deal of a dictum that "there is no mental differene without a physical difference," but he admits that this is a bit vague. Nonetheless, he does dispense with minds in favor of bodies, or, in the more refined version, of space-time regions. But "note that the situation is not symmetrical. The converse move of dispensing with bodies in favor of minds is not open to us, for we would not allow that there is no physical difference in the world without a mental difference — not unless we were idealists." This converse dictum is no doubt as vague as the original one, but, if we *were* idealists, we would seek to formulate it more carefully within a general metaphysical scheme. Where each body is an "embodiment" of mind, in some sense, and differences in physical states are suitably provided for in terms of embodiments together with "manifestations" of mind, some such dictum could no doubt be made sense of. Whether it would be assumed to hold or not, in a given idealist system, may be left open for the moment. But no matter, a theory of "embodiment" and "manifestation" must, it would seem, play a crucial role here.[11] The possibility of a critical, logical, metaphysical idealism remains an intriguing possibility, and one that can be harmonized with some of the desirable tenets of both logical positivism and scientific realism.[12] This is, however, a possibility that Quine "would not allow." Nor would he allow the development of the kinds of metaphysical idealism especially appropriate for accommodating the kinds of lofty values highly cherished by much of civilized mankind, East and West, down the centuries.[13]

A final comment. Let us reflect, a little more deeply than above, upon Quine's conception of "latter-day" physics. The eminent physicist J.A. Wheeler has pointed out that "Era I [in the history of physics] gave us the parabolas of Galileo and the ellipses of Kepler, motion [but] with no explanation of motion. Era II gave us the mechanics of Newton, the electrodynamics of Faraday and Maxwell, the geometrodynamics of Einstein and the chromodynamics of our day, law that explained change, but law with no explanation of law. Out of Era III, "meaning physics", we have to seek nothing less than the foundation of physical law itself."[14] Of this latter, he comments: "How can anyone even dream of abandoning for the foundation of existence a physics hardware located "out there" and putting in its stead a "meaning" software located who knows where? And how can the hard structure of hard science be moved over, solid as ever, onto this new and other-worldly foundation? Yet, despite all difficulties, and they are great, that must be the task and achievement of the coming Era III of physics." Quine's conception at best seems to be of a physics somewhere in the middle of Era II, but with quantum mechanics with its attendant "software" conveniently set aside to avoid complications.

Wheeler's choice of 'meaning' to help designate the coming era is perhaps unfortunate. He takes the word in the sense of Føllesdal, that "meaning is the joint product of all the evidence that is available to all who communicate."[15] Much depends here upon how 'evidence' is construed, upon how conflicts of evidence are handled, upon how evidence is related to what is evidenced, and so on. Also 'meaning' is taken here as a grand term — it should really be capitalized — for all meanings past, present, future, here, there, everywhere, so that 'the meaning of' as applied to specific words and phrases is given no account. For quantum theory, it would seem, this grand construal of 'meaning' could be replaced by some narrower notion that can readily be handled within the kind of language-system required. What sort of a notion could this be?

In his paper given at the Fifth International Congress for Logic,

Methodology, and Philosophy of Science[16], Wheeler made a good deal of the notion of *observership*, which seems more specific for the intended purpose. Of course 'observership' suggests an observer, and the logic of the term would seem to require this. Let 'Obs' be the predicate for observing as significant in the context

(1) '*p* Obs *x*,*a*',

expressing that observer *p* observes "entity" *x* as taken under the one-place predicate-description *a*.[17] The variable '*x*' here is to range over whatever the entities of the theory are and '*a*' over the one-place predicates constructed out of its primitive terms. It might be thought that the form (1) would give us all that is needed. But no. The immediate *acts* of observing, the states of a proton as observed, and so on, must be provided for. The event- or process-predicate '<Obs>' is needed for this. Let

'<Obs> *e*',

then, express that *e* is an act of event of *observing* of the kind quantum theory needs.

In place of (1) we could then introduce — in familiar fashion in process- or event-logic —

'*p* Obs_e *x*,*a*' as short for '(<Obs> *e* . *p* Instr *e* . *e* $\mathrm{Of}_{\mathrm{Object}}$ *x* . *x* Under_e *a*)',

where Instr is the relation of being the instrument — or agent — of the observing, *x* the object, and *a* the predicate-description giving the result of the observing (or the predicate-description *under* which *a* is taken).[18] If this definition — or something like it — is feasible, the notion '<Obs>' can be incorporated within the language of quantum theory. Of course, to do this in detail would not be easy. In principle, however, to do so would seem eminently desirable. We would have to show how its use would help to *précisé* the formulation of statements about the elementary quantum phenomenan, about complementarity, position, momentum, the wave-function, Bell's inequality, Wigner's notion of "amplification," probability amplitudes, and so on and on.

The question also arises as to whether '<Obs>', or something like it, is needed for a full schematization of the physics of Eras I and II. And even Quine's view can be stated only by making fundamental use of acts of *identifying* and *explaining*: "By identifying each space-time point with a quadruple of real or complex numbers according to an arbitrary system of coordinates, we can explain the space-time regions as sets of quadruples of numbers." Our acts of identifying and explaining, here, are fundamental to Quine's formulation, but do not therewith transform space-time points into quadruples of numbers as a fact of the matter. It is to be feared that a full statement of Quine's view — equations-*cum*-"text" — would require also the resources of Era II physics.

In attempting to *précisé* the language of Era II physics, we must be careful to distinguish the roles of '<Obs>', say, from that of (some representation of) 'consciousness', 'knowledge', 'information', and so on. Physicists tend to use these indiscriminately. Thus Eugene Wigner, for example, has commented that "physicists have found it impossible to give a satisfactory description of atomic phenomena without reference to consciousness."[19] And Heisenberg has commented that "the laws of nature which we formulate mathematically in quantum theory deal no longer with the elementary particles themselves but with our knowledge of this behavior."[20] Many examples of such statements from eminent physicists could be given. The point here is not to call into question their general intent. Strictly, however, a logical reconstruction of the language of quantum mechanics as such would presumably not make use of the whole gamut of epistemic words associated with 'mind', 'thought', 'knowledge', and so on. If the suggestion above is feasible, we should be able to get along with just '<Obs> *e*' and the full theory needed concerning it. A full epistemics would interanimate this notion with the others. Quantum mechanics, however, would seem to be a somewhat narrower enterprise, but still a much richer one than can be envisaged within Quine's narrow confines.

CHAPTER EIGHTEEN

On collective terms, mereological sums, and quantifiers

"Quae amissa salva."

It has often been pointed out that the collective use of general terms may conveniently be handled in terms of mereological sums. Bertrand Russell's remark

(1) 'Americans are the unhappiest of people'

does not lend itself easily to an analysis by means of quantifiers. It is not each and every American who is said here to be unhappy. No, the term 'Americans' is to be construed collectively and the remark ascribes unhappiness to Americans as a whole or as a collective. It is the mereological sum

(2) $(p_1 \cup p_2 \cup \ldots)$,

where p_1, p_2, etc., is the whole list of Americans to whom Russell intends reference, that is ascribed unhappiness in the way unhappiness is ascribable to collectives rather than to individual persons. A collective of people cannot be said to be unhappy in precisely the same sense that an individual can be. The two meanings are interrelated, of course, and it is one of the functions of a theory of hedonics to say precisely how.

The sum (2) is spelled out above enumeratively, but in practice this is rarely done. In fact, in many instances it may not be possible to do so. Instead, where 'Amer p' expresses that p is an American, in just the sense intended, we need a notation to give us the sum required. This sum is often referred to as the *fusion* of (the class of) Americans, where — recall the definition of 'FuPer p' in XV above —

(3) 'FuAmer p' could abbreviate '$((\mathrm{E}q(\mathrm{Amer}\ q\ .\ q\ \mathrm{P}\ p)\ .\ (x)(x\ \mathrm{P}\ p \supset (\mathrm{E}y)(\mathrm{E}q)(y\ \mathrm{P}\ x\ .\ y\ \mathrm{P}\ q\ .\ \mathrm{Amer}\ q\ .\ q\ \mathrm{P}\ p)))$'.

The fusion p of all Americans is then the one merological sum containing at least one American as a summand and such that every part of p has a part that is a part of some American that is a part of p.

We could go on to speak of fusions of classes, or even of virtual classes, but the phrase 'mereological sum' seems more direct and more suggestive of what it purports to name. In general, then, let us speak of a mereological sum of entities x of which such and such holds, or satisfying such and such a condition. Let us let

(4) '$(x\sigma$--x--$)$'

be significant as a primitive term, where '--x--' is a sentential form containing 'x' as a free variable and possibly free occurrences of other variables as well. (4) is to name the (generalized) mereological sum of all entities satisfying the condition that --x--.

Note that (4) is akin to the expressions of the form

'$(x1$--x--$)$'

used in V above. These expressions are only for sums of *unit* entities. (4) is a much more general form, however, not confined just to units. The two forms are closely intertwined however, as we see by nothing that

'$(x\sigma$--x--$) = (y1(\mathrm{E}x)(y \ \mathrm{P} \ x \ . $ --x--$))$'

and

'$(xl$--x--$) = (x\sigma(\mathrm{Unit} \ x \ . $ --x--$))$'

would hold as theorems in any language containing both types of expressions and having '--x--' as a sentential form in common.

Note that in the definition (3) fundamental use is made of the quantifiers occurring in the definiens, and such a definition is unthinkable without them. If we take expressions of the form (4) as primitive terms of a new theory, we may provide for the quantifiers definitionally as well as for all the Boolean or mereological notions of the calculus of individuals. Let us show how this may be done by reflecting again upon the systems of V, but with the quantifiers dropped. The resulting system is of interest on its own account, and not only as providing a very simple and straightforward way of handling the collective uses of general terms.[1]

Our primitive vocabulary, then, consists of 'P' for the part-whole relation, 'Ord' for the relation of ordinating, and '$\sim$' and 'v' as the primitive truth-functional connectives. In addition, all terms of the form (4) are primitives, where '--x--' is a sentential form of the language.

Let us note that '.', '$\supset$', and '$\equiv$' are immediately definable in familiar ways, and also that

D1. '$x = y$' is short for '(x P y . y P x)'.

Also we have the two very immediate abbreviations, of

D2. 'W' for '($x\sigma x$ P x)',

and of

D3. 'N' for '($x\sigma\sim x$ P x)'

D1 gives us identity, *D2* the universal or world element, and *D3* the null one.

With 'P' and 'W' available, we may define the universal and existential quantifiers in context by letting first

D4. '(x)--x--' abbreviate '(W P ($x\sigma$--x--) . --N--)',

and then, in familiar fashion

D5. '(Ex)' may abbreviate '$\sim$ (x) $\sim$'.

Allowance is made here, it should be observed, for the quantifers to cover the null individual.[2]

The remaining Boolean operators, together with relations of overlapping and of being discrete, are now also definable.

D6. '($x \cup y$)' for '($z\sigma$(z P x v z P y))',
D7. '($x \cap y$)' for '($z\sigma$(z P x . z P y))',
D8. 'x O y' for '(Ez)($\sim$ z P N . z P x . z P y))',
D9. 'x D y' for '$\sim$ x O y',

and

D10. '-x' for '($y\sigma y$ D x)'.

D6 introduces ordinary sums, *D7* ordinary Boolean products, *D8* the relation of overlapping, *D9* that of discreteness, and *D10* the Boolean negation.

As a few basic principles of the theory we should note the following.

Pr1. $\vdash x$ P x,

Pr2. $\vdash$ (x P y . y P z) $\supset$ x P z.

Also we would have that

Pr3. $\vdash$ --x-- $\supset$ x P ($y\sigma$--y--), where '--x--' differs from '--y--' only in containing free occurrences of 'x' wherever there are free occurrences of 'y' in '--y--',

Pr4. $\vdash$ ($x\sigma$--x--) P y $\equiv$ W P ($x\sigma$(--x-- $\supset$ x P y)), where 'y' does not occur freely in '--x--',

Pr5. $\vdash$ x P ($y\sigma$--y--) $\equiv$ W P ($y\sigma$(y P x $\supset$ $\sim$ (z σ $\sim$ ($w\sigma$(z P y . z P w . --w--)) P N)) P N), where 'y' and 'z' do not occur freely in '--w--' and '--w--' and '--y--' differ from each other appropriately,

Pr6. $\vdash$ (W P ($x\sigma$--x--) . --N--) $\supset$ --y--, where (etc.),

Pr7. $\vdash$ (W P ($x\sigma$(--x-- v .....)) . (--N-- v .....) . $\sim$ .....) $\supset$ W P ($x\sigma$--x--), where '.....' is a formula not containing 'x' as a free variable,

and that

Pr8. $\vdash$ ($x\sigma$--x--) = ($x\sigma$($\sim$ x P N . --x--).

Pr1-*Pr2* give us two basic properties of P, and *Pr3*-*Pr8* the same basic properties of generalized mereological sums. Presupposed, of course are principles yielding the truth-functional tautologies plus the rule of Modus Ponens. It should be noted that from *Pr6*, in view of *D4*, we can prove that

(5) $\vdash$ (x)-x-- $\supset$ --y--, where (as needed).

Similarly we have from *Pr7* that

(6) $\vdash$ (--x-- v .....) $\supset$ ((x)--x-- v .....), where (as needed).

Also we must have a *Rule of Generalization* of the form

Gen. If $\vdash$ --x-- then $\vdash$ (W P ($x\sigma$--x--) . --N--),

which of course gives us immediately by *D5* that

Gen'. If $\vdash$ --x-- then $\vdash$ (x)--x--.

It is well known that *Gen'* together with (5) and (6) give us a viable rendition of the standard first-order quantification-theory as developed with the help of free variables.

In the foregoing, fundamental use has been made of free variables. If these are to be dispensed with, we would proceed a little differently. In place of *Pr1* we would have that

Pr1'. $\vdash$ W P ($x\sigma x$ P x),

providing that every element is a part of itself. Let now

D11a. '(xyσ--x--y--)' abbreviate '(xσW P (yσ--x--y--))',

D11b. '(xyzσ--x--y--z--)' abbreviate '(xyσW P (zσ--x--y--z--))',

and so on. Then we would assume, in place of *Pr2*, that

Pr2'. $\vdash$ W P (xyzσ(x P y . y P z) $\supset$ x P z)),

providing for the transitivity law for P. And similarly for the other principles providing the other basic properties of P and of the mereological sums. And still further principles would be needed to assure that all laws of quantification are forthcoming without the use of free variables.

To provide for the theory ordinal individuals, further principles akin to those of V are of course needed, no matter how the quantifiers are handled. If any non-logical individual constants for specific cardinal or ordinal individuals (including collectives, attributive relations, relations of higher degree, as well as relation-individuals of higher logical type) are taken as primitives, appropriate allowance must be made for such in the statement of the laws and rules. And of course, suitable existence principles, as in V, are needed all along the line.

The foregoing is not intended to provide a full formalization of the theory of generalized mereological sums, but only a rough sketch. It is not necessary to develop here an axiomatization or to exhibit the multitude of interesting theorems that would result. Such an enterprise would require a considerable technical development not needed for present purposes. The aim here has been merely to show that such a development is possible and to exhibit a few principles that should obtain.

There are of course many alternative ways of providing for the quantifiers currently on the market. Some make use of both free and bound variables provided for primitively[3]; some of only bound variables[4]; some of only free variables but with bound variables allowed to form abstracts[5] or special terms such as the Hilbertian ε-expressions[6], or the like; some without variables at all, such as in predicate functor logic, relational algebra, relational-term logic, combinatory logic, and the like, as noted in XIII above. All of these methods have interesting features and good reasons are forthcoming for espousing them. The use of the expressions for generalized mereological sums above provides one more item in this arsenal of methods. It helps us to see that collective terms are quite as fundamental as distributive ones, and in fact the latter can be handled in terms of the former in view of *D4* and *D5*. To say, for example, in virtual-class terms that

$$(x)(\{y \ni \text{--}y\text{--}\}\, x \supset \{y \ni \ldots y \ldots\}\, x),$$

construed distributively, is merely to make a certain statement about the mereological sum

$$(x\sigma(\text{--}x\text{--} \supset \ldots x \ldots))$$

and about N.

Let us reflect for a moment upon some collective terms in English, as discussed in Fowler.[7] "Fowler", according to Geach, "clearly had no proper logical education, but he shows a good nose for logical points."[8] A good nose probably best reveals itself in being able to sniff out a reasonably good preliminary classification. Under 'collectives' Fowler writes: "The word is applied to many different things. What is common to all is that a noun singular in form is used with a plural implication.... Collectives may be roughly divided into the following groups. 1. Nouns denoting a whole made up of similar parts.... These are also called *nouns of multitude*. 2. Nouns that make no plural form but are used as both singular and plural.... 3. Nouns that make plural forms in the normal way but whose singular form may also be used as a plural, sometimes with little if any difference of meaning.... 4. Names of materials used for a collection of things made from them.... 5. Words of number or amount that when used after definite or indefinite numerals have the singular rather than the plural form.... 6. Abstract singulars used instead of concrete plurals.... 7. Even nouns denoting substances of indefinite quantity such as *butter* and *water* are classed as collectives."

Note that of these seven groups, only five are characterized semantically in terms of 'denotes' or 'names' or as explicitly denoted or named. We end up than with collective terms characterized as (1) nouns of multitude, (4) names of materials, (5) names of number or amount (6) abstract nouns, and (7) names of substances of indefinite quantity. Examples of (1) are 'committee', 'firm', 'orchestra', 'soviet', of (4) 'china', 'linen', 'silver', of (5) 'dozen', 'score', 'hundred', of (6) 'accommodation' (in the sense of 'rooms or lodgings'), 'kindling', 'royalty', 'pottery', of (7) 'butter' and 'water'. Sentences containing these words may be given appropriate logical forms in terms of mereological sums.[9] And, indeed, there appear to be no ways of doing this without the use of such sums.

CHAPTER NINETEEN

On the effectiveness of mathematics in the natural sciences: Wigner, Whitehead, and Carnap

"Quae non valeant singula, juncta juvant."

In a famous paper[1], the eminent physicist and Nobel lauriate Eugene Wigner has commented "that the enormous usefulness of mathematics is something bordering on the mysterious and that there is no rational explanation for it." Further, "it is just this uncanny usefulness of mathematical concepts that raises the question of the uniqueness of our physical theories." He ends his discussion with the cheerful observation that "the miracle of the appropriateness of the language of mathematics for the formulation of the laws of physics is a wonderful gift which we neither understand nor deserve. We should be grateful for it and hope that it will remain valid in future research and that it will extend, for better or for worse, to our pleasure, even though perhaps also to our bafflement, to wide branches of learning." As to this last, there need be no fear, as we shall try to show. And of course the gift of mathematics, along with much else in our human life, is a wonderful gift. And, no doubt, the appropriateness of mathematics for whatever it touches is a "miracle," along with the basic miracle of being or existence itself — to recall a notion due to Paul Tillich. For the present, let us leave the mystery, miracles, and gifts, *as such*, aside in order to ponder the reasonableness of them and hence to arrive at some modicum of philosophical understanding of them.

Let us begin by reflecting, one by one, upon the items Wigner raises. Borrowing from Walter Dubislav, he comments that "somebody ... said that philosophy is the misuse of a terminology invented for just this purpose. In the same vein, I would say that mathematics is the science of skillful operations with concepts and rules invented for just this purpose. The principal emphasis is on the invention of concepts. Mathematics would soon run out of interesting theorems if these had to be formulated in terms of the concepts which already appear in the axioms." Clearly, now, *mathematicians* are skillfull at operating with certain concepts or notions. Perhaps they "invent" them; perhaps they "discover" them. If we take the former view, we have the difficult task of explaining why what is invented is so "objective" and compelling and exerts such coercion upon all minds sufficiently trained to understand it. Once grasped, mathematical truths are of universal validity. Even so, it seems somewhat perverse to think that mathematical notions are "invented" or discovered just for the purpose of skillfully operating with them. By parity of phraseology, one would have to say that the great Bach fuges were written just for the purpose of playing them skillfully. Any such characterization would be one-sided and would leave the real heart of the matter out of account.

We would hope, as we read on, that Wigner's account of mathematics would make up for this one-sidedness, but in this we are disappointed. Consider, for example, the comment that "the rules for operations with pairs of numbers are obviously designed to give the same results with fractions which we first learned without reference to 'pairs of number.'" Let us be sure to distinguish here the contingency of how and what we happen to learn from the objectivity of what it is that is learned. Let us realize also that the logic of ordered pairs of numbers, or of pairs of anything else, is quite independent of any talk of fractions and owes its "validity" to the underlying logical principles governing them and not to those governing fractions. There would be ordered pairs even if there were no fractions. The laws governing ordered pairs are thus not "obviously designed" just to accommodate the fractions. How is it in fact that they do accommodate them so well? Because the notion of an ordered pair is of such vast generality as to allow all manner of "entities" to be components of a pair. The "reason" for this latter is that it is an instance of a general principle implicit in the material of V above, this instance being that

$$\vdash (x)(y)((\mathrm{Nn}\ x\ .\ \mathrm{Nn}\ y) \supset (\mathrm{E}z)z\ \mathrm{Ord}\ x,y),$$

that any natural numbers x and y are such that there exists a z that bears Ord to them.

As a second example, Wigner notes that the "rules for the operations with sequences, that is, with irrational numbers, still belong to the category of rules which were determined so as to reproduce rules for the operations with quantities which were already known to us." Here, too, we should distinguish the general logic of sequences from the particular principles needed for handling sequences of numbers, and both of these from what happens to have been "already known to us." The logic of sequences, a part of the general theory of relations[2], is a subject of immense importance and interest quite independent of its providing the foundation for the arithmetic of real numbers. It is surely a distortion to say that it was "determined so as to reproduce rules for the operations with quantities already known to us." It was determined by the discovery and explicit formulation of general logical laws of universal validity concerning dyadic relations that are connected, transitive, and contained in diversity.[3] We shall try to show later just why such laws are helpful for gaining conceptual clarification in physics as well as elsewhere.

Wigner writes that, "whereas it is unquestionably true that the concepts of elementary mathematics and particularly elementary geometry were formulated to describe entities which are directly suggested by the actual world, the same does not seem to be true of the more advanced concepts, in particular the concepts which play such an important role in physics.... Most more advanced mathematical concepts, such as complex numbers, algebras, linear operators, Borel sets — and this list could be continued almost indefinitely — were so devised that they are apt subjects on which the mathematician can demonstrate his ingenuity and sense of formal beauty.... The complex numbers provide a particularly striking example for the foregoing. Certainly, nothing in our experience suggests the introduction of these quantities. Indeed, if a mathematician is asked to justify his interest in complex number, he will point, with some indignation, to the many beautiful theorems in the theory of equations, of power series, and of analytic functions in general, which owe their origin to the introduction of complex numbers. The mathematician is not willing to give up his interest in these most beautiful accomplishments of his genius."

Let us not deprive the mathematician of his interest in formal beauty, and there can be no doubt but that the long, slow, almost 2000 year development leading up to the theory of functions of a complex variable is one of the greatest achievements of the human mind in the whole of history. The reasons for this beauty and greatness are not so difficult to fathom as Wigner would have us believe. He omits to mention a crucial item concerning the fundamental arithmetic operations of addition, subtraction, multiplication, division, exponentiation, and taking the roots of. In the system of the positive integers, we have the closure principle for addition, that, given any two integers, the sum of them is an integer. We have also a closure principle for multiplication and exponentiation, but not fur subtraction, division, or taking the roots of. A closure principle is a thing of beauty and a desire forever, and the development of the signed integers, of the positive and of the signed rationals, of the real numbers, and of the complex numbers was motivated, in part surely, by the desire to have closure principles — even if with certain restrictions — available for all of these fundamental operations. There is thus, indeed, something in mathematical experience, at each step of this development, to suggest the introduction of the new kind of number.

Wigner makes much of the "invention of concepts," as we have already noted, and contends that "mathematics would soon run out of interesting theorems if these had to be formulated in terms of the concepts which already appear in the axioms." Here we should distinguish between primitive and defined concepts, and between a "pure" mathematics and an "applied" one. The axioms need not all be stated just in primitive terms, although they could be if desired. Ordinarily some defined notions occur in the axioms along with the primitive ones. As the system is developed, more and more definitions are introduced and therewith more and more "new" concepts. It is often surprizing that such concepts become available in this way, and that "interesting" theorems are provable (from the axioms) concerning them. In a pure mathematics, all the definable notions must be regarded as mathematical ones. In an applied mathematics, there must always be at least one non-mathematical primitive notion, usually a predicate, applicable to the non-mathematical objects under consideration. Concerning this predicate or these predicates, suitable non-mathematical axioms are given laying down the salient non-mathematical principles needed. But whatever we do here, if the system itself is of any genuine usefulness, there are always interesting theorems forthcoming. And ultimately, if all definitional abbreviations are eliminated, these may all be stated wholly in terms of the primitives, which already appear in the axioms.

Mention of applied mathematics leads us at once to the portals of physics. "The physicist is interested in discovering the laws of inanimate nature," we are told, and such laws "contain, even in their remotest consequences, only a small part of our knowledge of the inanimate world." As philosophers, we seek also to *understand*, or to make sense of, the statements in which these laws are expressed, to see how they relate one to another, and to incorporate as many as we can within a systematic framework. The endeavor to understand, in this sense, may also serve as a heuristic for the discovery of new laws. Discovery and systematization are thus not to be thought of as separate enterprises; each may contribute to the other in fundamental ways.

What, now, is the role of mathematics in physical theories? "Naturally," Wigner observes, "we do use mathematics in everyday physics to evaluate the results of the laws of nature [which are always of conditional form], to apply the conditional statements to the particular conditions which happen to prevail or which happen to interest us. In order that this be possible, the laws of nature must already be formulated in mathematical language." For this remark to be cogent, we should construe 'mathematics' in a somewhat broader sense than Wigner intends. It should include also the Bridgmanian "text" (already referred to in XVII above), or rather, the text in a semiotically formalized format. The laws of nature are to be stated, not just in the language of pure mathematics but in the "mixed" language incorporating the text as well. And herein lies the difficulty. The languages of logic and pure mathematics are already formulated, but that embodying the text ordinarily is not. Wigner notes that "there are two basic concepts in quantum mechanics: states and observables. The states are vectors in Hilbert space, the observables self-adjoint operators on these vectors." This, of course, is only a part of the story, a full formalization of the whole text being needed before anything even remotely accommodating quantum mechanics in any intelligible sense is achieved.

We are now in a position to attempt to answer — to some extent, anyhow — the question as to why logic and mathematics play so important and effective a rule in the natural sciences. For this, let us construe 'logic' in a wide sense so as to embrace mathematics as formulated, say, in the manner of V above, as well as a semiotics or metalogic, as essentially in II above, to provide for any relevant text. Logic may then be conceived as the most general theory of mereology and/or "ontology" needed for natural language. Special scientific, technical terms may be added as new primitives as needed. The means are then at hand for expressing, in a concrete linguistic form, whatever is desired in a given context. This is the sense in which the laws of nature may be said to be "already ... formulated in mathematical language" — not explicitly written down, but formulable in terms of the linguistic resources at hand.

The connection between a formulated natural science and the underlying ontological-semiotical framework is that between *exempla* and the underlying *principia*, between a specialized instance and a completely general theory containing, in a certain sense, all possible instances. No account short of this would seem to shed much philosophical understanding on this complex interrelationship. In Wigner's view, the connection seems "unreasonable," bordering on the mysterious, and lacking in rationality. The real mystery here concern the nature of logic itself, our main source of rationality, without which mathematics, the sciences, and all serious discourse crumble to the dust.

In another famous essay[4], Whitehead, in speaking of algebra as the study of "patterns," comments that "mathematics is the most powerful technique for the understanding of pattern, and for the analysis of relationships of pattern.... Having regard to the immensity of its subject-matter, mathematics, even modern mathematics, is a science in its babyhood. If civilization continues to advance, in the next two thousand years the overwhelming novelty in human thought will be the dominance of mathematical understanding." Whitehead does not confine mathematics here just to the study of geometry and number, but regards it as the study of pattern in general. And by a pattern here we can mean a relational structure in the sense of the modern logic of relations.[5] Relations — to reiterate somewhat — are essential in delineating structures of sufficient complexity for mathematics and the sciences, and *a fortiori* for any systematic discourse in the humanities.[6]

Earlier in his essay, incidentally, Whitehead has some rather harsh things to say about the theory of number as based upon the theory of types. Because the formulation of numerical mathematics presupposed in this present paper contains a version of type-theory, as noted in V above, it will not be out of place to examine the points Whitehead raises in order to show that they are not applicable here.

"Russell introduced the notion of "types" of entities," Whitehead writes. "According to that doctrine, the notion of number should only be applied to a group of entities of the same type. Thus the number "three" as applied to entities of one type has a different meaning to the number "three" as applied to entities of another type. For example, if we are considering two different types, there are two different meanings of the number "three." Russell was perfectly correct. By confining numerical reasoning within one type, all the difficulties [contradictions or paradoxes] are avoided. He had discoverd a rule of safety. But unfortunately this rule cannot be expressed apart from the presupposition that the notion of number applies beyond the limitations of the rule. For the number "three" in each type, itself belongs to different types. Also each type is itself of a distinct type from other types. Thus, according to the rule, the conception of two different types is nonsense, and the conception of two different meanings of the number three is nonsense. It follows that our only way of understanding the rule is nonsense."

Let us broaden somewhat the notion of natural number as sketched in OM. The only entities assigned numbers there are the cardinal ones, although the numbers themselves are construed as infinite logical sums of certain kinds of ordinal entities. Let us put forward now a new notion of number, by broadening the key relation Sm of equinumerosity or similarity. For this, also, we must broaden the definition of

'$x\ u_2\ y$' to be now short for '$((z)((\mathrm{Unit}\ z\ .\ z\ \mathrm{P}\ u) \supset (\mathrm{E}w)(\mathrm{E}v)z = (w\ \mathrm{ord}\ v))\ .\ (x\ \mathrm{ord}\ y)\ \mathrm{P}\ u)$',

so that u_2 here is a dyadic relation-individual between any x and y and not just between cardinal entities. In OM, Sm is introduced as a relation significant only between cardinal entities. Let us weaken this requirement somewhat and let

'x Sm y' abbreviate now '$(\mathrm{E}w_2)((1{\rightarrow}1)w_2\ .\ (z)(u)(z\ w_2\ u \supset (\mathrm{Unit}\ z\ .\ \mathrm{CardInd}\ z\ .\ z\ \mathrm{P}\ x\ .\ \mathrm{Unit}\ u\ .\ u\ \mathrm{P}\ y))\ .\ (z)((\mathrm{Unit}\ z\ .\ \mathrm{CardInd}\ z\ .\ z\ \mathrm{P}\ x) \supset (\mathrm{E}u)(\mathrm{Unit}\ u\ .\ u\ \mathrm{P}\ y\ .\ z\ w_2\ u))\ .\ (u)((\mathrm{Unit}\ u\ .\ u\ \mathrm{P}\ y) \supset (\mathrm{E}z)(\mathrm{Unit}\ z\ .\ \mathrm{CardInd}\ z\ .\ z\ \mathrm{P}\ x\ .\ z\ w_2\ u)))$'.

Thus x and y are similar if there is a one-one relation between their respective unit parts. Letting now

'Nc'x' abbreviate '$(y1(\mathrm{E}z)(y\ \mathrm{Ord}\ z,z\ .\ \mathrm{CardInd}\ z\ .\ z\ \mathrm{Sm}\ x))$',

we arrive at a notion of the cardinal number of x no matter what kind of an entity x may be. Also independently of this broadened relation Sm, we may let

'1' abbreviate '$(x1(\mathrm{E}y)(x\ \mathrm{Ord}\ y,y\ .\ \mathrm{CardInd}\ y\ .\ \mathrm{Unit}\ y))$',

'2' abbreviate '$(x1(\mathrm{E}y)(\mathrm{E}z)(x\ \mathrm{Ord}\ (y \cup z),(y \cup z)\ .\ \mathrm{Unit}\ y\ .\ \mathrm{Unit}\ z\ .\ \mathrm{CardInd}\ y\ .\ \mathrm{CardInd}\ z\ .\ {\sim}\ y\ \mathrm{P}\ z))$',

and so on. The required properties of the natural numbers are then forthcoming in the system, on the basis, of course, of appropriate axioms.

Note that the cardinal number of x is taken here as the sum of dyadic ordinal individuals both of whose ordinands are the same cardinal entity, this entity bearing Sm to *x*. Why not take it merely as the sum of all entities, cardinal or otherwise, bearing Sm to *x*? If we did this latter, we would have the unfortunate consequence that 1 would equal 2. 1 would then become the sum of all unit entities, and thus W itself. 2 would become the sum of all dual entities, and thus W itself! The trick of the method used here is to require the notion of being a *unit part of* to play the erstwhile role of being a *member of* in set- or class-theory. To be a unit part of 2 here is to be a unit *x* determined by a dual entity ($x \cup z$), the way of determining this being to require that *x* Ord ($y \cup z$), ($y \cup z$). If 2 were the mere sum of all duals, its unit parts would not be parts determined in some way by dual entities. The use of sums of *unit parts* — both cardinal and ordinal — is fundamental to the method here and enables us to avoid all reference to sets or classes.

The method of introducing numbers here contrasts with that presented in the preliminary sketch of OM, in the Appendix below, which makes fundamental use of virtual classes and relations. The method here, on the contrary, avoids all such use, and is allied to the second method of OM (not mentioned in the Appendix). There are still further ways of introducing cardinal numbers, on the basis of either method, that are worth exploring. It is far from clear yet that there is any one method — even in set-theory — that is fully satisfactory from all points of view.

Let us observe now that Whitehead's objections to the theory of types do not apply to the theory of cardinal numbers just sketched. Let us consider the one number 3, as introduced by the definition that

> '3' abbreviates '(*x*1(E*y*)(E*z*)(E*w*)(*x* Ord ($y \cup z \cup w$),($y \cup z \cup w$) . Unit *y* . Unit *z* . Unit *w* . CardInt *y* . CardInd *z* . CardInd *w* . ~ *y* P *z* . ~ *y* P *w* . ~ *z* P *w*))'.

According to this, 3 is the infinite logical sum of all dyadic ordinal individuals whose two ordinands are some one and the same sum of three cardinal units.

In Russell's theory, Whitehead observes, "if we are considering two types [of entities], there are two different meanings of the number "three"." Let a, b, and c by any three distinct cardinal units, say, and u_2^1, u_2^2, u_2^3 any three distinct unit dyadic relations between cardinal entities. Clearly we have then that

$$\text{Nc'}(a \cup b \cup c) = 3,$$

and also that

$$\mathrm{Nc}'(\mathrm{u}_2^1 \cup \mathrm{u}_2^2 \cup \mathrm{u}_2^3) = 3.$$

Both of these depend upon the principles that

$$\vdash (x)(y)(z)((\mathrm{Unit}\ x\ .\ \mathrm{Unit}\ y\ .\ \mathrm{Unit}\ z\ .\ \sim x\ \mathrm{P}\ y\ .\ \sim x\ \mathrm{P}\ z\ .\ \sim y\ \mathrm{P}\ z) \supset (w1(\mathrm{E}u)(w\ \mathrm{Ord}\ u,u\ .\ u\ \mathrm{Sm}\ (x \cup y \cup z))) = (w1(\mathrm{E}x')(\mathrm{E}y')(\mathrm{E}z')(\mathrm{Unit}\ x'\ .\ \mathrm{Unit}\ y'\ .\ \mathrm{Unit}\ z'\ .\ \mathrm{CardInd}\ x'\ .\ \mathrm{CardInd}\ y'\ .\ \mathrm{CardInd}\ z'\ .\ \sim x'\ \mathrm{P}\ y'\ .\ \sim x'\ \mathrm{P}\ z'\ .\ \sim y'\ \mathrm{P}\ z\sim'\ .\ w\ \mathrm{Ord}\ (x' \cup y' \cup z'),(x' \cup y' \cup z')))).$$

Whitehead also objects that "the number "three" in each type, itself belongs to different types," in Russell's account. In the theory here, however, 3 is of just one type but can be the cardinal number of any entity having as summands dyadic ordinal entities whose ordinands are any triples of distinct units whatsoever. "Also each type is itself of a distinct type from other types. Thus, according to the rule, the conception of two different types is nonsense,..." But not here. Let V_1 be the "type," so to speak, of cardinal entities, and V_2^2 of attributive relations of cardinal entities. We may let

'u_2 Att x' be short for '$(\sim u\ \mathrm{P}\ \mathrm{N}\ .\ (y)((\mathrm{Unit}\ y\ .\ y\ \mathrm{P}\ u) \supset (\mathrm{E}z)y\ \mathrm{Ord}\ z,z)\ .\ (x\ \mathrm{ord}\ x)\ \mathrm{P}\ u\ .\ (z)(w)((z\ \mathrm{ord}\ w)\ \mathrm{P}\ u \supset z = w))$',

expressing that u_2 is an attributive relation of x no matter what the type of x is. Then, also, we may let

'$\{z{:}(z = x \vee z = y)\}$' abbreviate '$(\iota u_2\ .\ (u_2\ \mathrm{Att}\ x\ .\ u_2\ \mathrm{Att}\ y\ .\ (z)(u_2\ \mathrm{Att}\ z \supset (z = x \vee z = y))))$'.

then, clearly,

$$\mathrm{Nc}'\ \{z{:}(z = \mathrm{V}_1 \vee z = \mathrm{V}_2^2)\} = 2;$$

And similarly for any other two distinct "types."

Note here that the use of '$\{z{:}(z = x \vee z = y)\}$' is an adaptation of the familiar notation for class-abstraction. Here it is taken as an expression for the attributive relation of applying to either x or y. In any full development of the system into the foundations of mathematics, such a notation, as well as one for relational abstraction in general, would be very useful. Rather than as classes of classes that correspond one-to-one with a given class (as with Whitehead and Russell), the cardinal numbers here are taken as infinite sums of attributive relations of cardinal units, of cardinal duals, and so on. The restriction to cardinal units, duals, and so on, here is to use as counters, so to speak, the simplest kinds of entities available in the system.

Incidentally, Whitehead himself, in 1934, published a paper purporting also to avoid the inadequacies of Russell's method.[7] Unfortunately, this paper has been almost totally neglected by the Whitehead commentators. There is need for a careful appraisal of it that cannot be undertaken here. *Prima facie*, there seem to be some disconcerting features in the system put forward that would require a very considerable discussion. Fundamental use is made of unanalyzed propositions *sui generis* together with quantifiers upon them. At best such entities are obscure and cry out for a clear-cut semantical characterization. Also '=' is taken as a primitive as between propositons, a notoriously difficult and controversial notion in need of much explication. In the interpretation of '$p = q$' much must depend upon the inner character of p and q. Unanalyzed propositional identity is a highly obscure notion on which to attempt to found mathematics. As a primitive, also, there is 'Ec!' in the context 'Ec!x' expression "a proposition about x ... which ascribes to x no intension other than that intension derived from purely logical notions. It is a proposition involving x in the extremity of abstraction from intension. This is what is meant by saying that it is a propositon involving x in pure extension. It expresses that recognition of individuality which is involved in counting." These comments need a good deal of further elaboration before any clear semantics for 'Ec!x' can emerge. Whitehead construes the unit class of x merely as the proposition Ec!x, therewith blurring the assertive or sentential character of a proposition with the entitival character of a unit class. And similarly for relations. Perhaps a fundamental category mistake is involved here. It is thus an open question as to whether a viable foundation for arithmetic is forthcoming in this paper. The matter needs a careful discussion, if only to place it properly within the whole Whitehead *corpus*.

Let us return now to our main theme.

In another important essay, presented at the Harvard Tercentenary[8], Carnap reflected upon logic under the general rubric "Factors Determining Human Behavior." "To see clearly on the matter," he writes, "it is essential that we obtain an adequate conception of the province of logic," and this latter "is concerned with *relations* between factual sentences" — and between logically true ones also, as well as between sentences of both kinds, it should be added. He goes on to point out "the requirements which thinking must satisfy in order to be logical or reasonable," namely, "clarity, consistency, and adequacy of evidence." Let us glance briefly at these by way of a needed supplement to Wigner's account.

Carnap discusses the condition of clarity by making much of the old positivist thesis of the meaningless of metaphysical sentences. It will be more to the point, for the present, to reflect upon the need for clarity in the language of science itself. 'Clarity' is a tricky word and can be made to cover much too much. Perhaps it should best be thought of in terms of a full delineation of relevant structure, both that between or among the sentences of the given language, and that which the various sentences themselves describe relationally of the objects the theory talks about. Concerning the former, we know now a good deal, thanks to the tremendous advances in modern logic. Concerning the latter, however, we know much less that we should like, thanks to the almost total neglect of these advances on the part of working scientists themselves. Quantum mechanics is a good case in point. The role of the observer, of the observations involved, and of the coding of these observations withing an appropriate symbolism, play a fundamental role. (Recall XVIII above.) A full logic of all this, which has not been supplied, would seem needed to provide for the semiotics of the "text" as well as to help us keep tabs on our tools. Whatever being "clear" about a given subject-matter is, surely explicit awareness of the relational structure involved and of the nature of the entities related is an essential ingredient. And the more detailed it is, the better. Like truth itself, *Gott wohnt im Detail*, as an old German adage has it.

The condition of consistency surely needs no further elaboration or justification at this late date. Without it, science itself crumbles. We are assuming here the usual two-valued logic of the truth-functional connectives, so that the law

(L) $\vdash \ulcorner(A \supset (\sim A \supset B))\urcorner$

holds. Thus if A and $\ulcorner\sim A\urcorner$ both hold, then any B does also, so that all sentences of the science then hold; the difference between truth and falsity collapses.

Concerning what is essentially this argument, Tarski has remarked, interestingly, that he "has some doubts whether ... [it] contains an adequate analysis of the situation. I think that people who do not know modern logic are as little inclined to accept an inconsistent theory as those who are thoroughly familiar with it; and probably this applies even to those who regard (as some still do) the logical law (L) on which the argument is based as a highly controversial issue, and almost as a paradox. I do not think that our attitude toward an inconsistent theory would change even if we decided for some reasons to weaken our system of logic so as to deprive ourselves of the possibility of deriving every sentence from any two contradictory sentences."[9] Tarski is not here condemning those who accept (L) as a logical law, but rather is concerned with (i) those who "do not know modern logic," and hence in particular the classical logic containing this law, and with (ii) those who accept an alternative kind of logic in which (L) does not figure as a law. If Tarski is correct concerning those in (i), this is splendid. For those under (ii), it should be observed that we know a great deal more now about deviant systems of logic than at the time Tarski wrote, and that there is much more evidence now, than then, that such logics presuppose the classical logic (containing (L) as a law) within the metalanguage in which the deviant logic is formulated. The law (L) need not occur within the deviant logic itself, but only within the semiotics in which the "text" concerning the logic is formulated.[10] Tarski's doubts here thus need no longer seriously detain us.

Carnap's third condition, that of adequacy of evidence, is not quite so easy. This condition is "that there be a sufficiently secure empirical foundation both for the universal laws we formulate and for the predictions we make with their aid." In the years since Carnap wrote this, it has become increasingly difficult to arrive at any satisfactory notion of adequacy of evidence. Also, emphasis upon the purely empirical character of evidence has been questioned, concepts and non-empirical features looming to be of greater import than Carnap indicates.

One more item concerning the effectiveness of mathematics in natural science should be mentioned before we close, namely the importance of *quantitative* concepts, without which the advanced sciences are unthinkable. Carnap, again, has skillfully called attention to these. "Quantitative concepts," he writes[11], "(also called metrical or numerical concepts or numerical functions) are those which serve for characterizing things or events or certain of their features by the ascription of numerical values; these values are found either directly by measurement or indirectly by calculation from other values of the same or other concepts.... Quantitative concepts are no doubt the most effective instruments in the scientific arsenal." Mathematics as the general study of relational structures does, after all, contain as a part the theory of quantity — let us not forget. This, together with the wherewithal needed for the formulation of the "text," give us two of the great pillars, surely, upon which the edifice of the sciences rests.

A final comment, in Wignerian style, concerning the gift of the myterious effectiveness of mathematics *überhaupt*. The mystery is that of rationality itself. Whence does it come and to what avail? Its gift is one of the great treasures of life, and logic is the means of its use. Rationality is the key to philosophic understanding. We are free to reject it if we wish, just as we are free to reject other treasures. The real bafflement would seem to be the widespread proclivity for such rejection. The treasures are ours, not just for the asking but only as the result of much hard work and stern discipline — intellectual, moral, and aesthetic. Unwillingness to undergo these, but in their stead to put something easier and more obvious, and more conductive to immediate earthly gain, seems to be characteristic of our human lot....

CHAPTER 20

On ordinal acts and era III mathematics

"Kein Weg; ins Unbetretene!"

In the original presentation of ordinal mereology, OM, it will be recalled, much was made of the predicate 'Ord' for the relation of ordination.[1] The form

'*x* Ord *y*,*z*'

expresses that *x* is the result of ordinating *y* and *z* in this order, in other words, that *x* is the ordinal individual consisting of *y* and *z* taken in this order. Ordinal individuals were assumed to be individuals *in rerum natura* on a part with the non-ordinal, cardinal individuals. It was remarked, rather casually, that 'Ord' is "the kind of predicate that gives rise [— naturally, we might add —] to an *event-descriptive predicate*.[2] Let

'<Ord> *e*'

express that *e* is an act or process of ordinating. The ordinal indivdual (*x* ord *y*)[3] is then the result of an act of ordinating *x* with *y*. Who or what is it that does the ordinating? It may be a human individual — the "creative subject" of intuitionistic mathematics (in one kind of rendition) — or a computer, or a social group of some kind, or whatever...." Let us take this casual remark seriously, now, to see whither it leads us. It will turn out to be a good deal more serious and interesting than was originally supposed.

Where <Ord>*e*, we may think of *e* as a kind of constructive *act* on the part of the logician or mathematician when doing his or her offical work. In fact, it may be thought of as the most fundamental kind of act he or she ever performs, for out of such acts the relations may be constructed that provide the wherewithal for doing mathematics. Let $\mathrm{On}_{\mathrm{Object}}$ and WithRespectTo be basic propositional relations[4], and let '*x* Result *e*' express that *x* is the *result* of, or results from, the act or process *e*. The *x* here might be a "concrete" object, but it need not be; it might be a state or even an act or process itself. In fact, the result could be taken as the very *e* such that <Ord>*e*, so that we may let

> '*e* Ord *x*,*y*' abbreviate '(<Ord>*e* . Act *e* . *e* $\mathrm{On}_{\mathrm{Object}}$ *x* . *e* WithRespectTo *y* . *e* Result *e*)'.

Here, even if *x* and *y* are "concrete" objects, or of whatever kind of object is regarded as acceptable in the given context, *e* may itself be taken as (*x* ord *y*).

We may go on to iterate Ord, as previously, so that

> '*e* Ord *x*,*y*,*z*' may abbreviate '(E*e*′)(*e*′ Ord *x*,*y* . *e* Ord *e*′,*z*)',

and so on. Although two styles of variables have been used here, we see that they reduce actually to just one. Any differentiation intended between the '*e*'s, etc., and the '*x*'s, etc., is provided by the context. By further iteration, all manner of ordinal individuals may be built up. These are to be the simplest, or "unit," ordinals.

We have seen previously how, in terms of 'Ord' together with the basic 'P' for a part-whole relation, we may go on to construct mathematical notions and to prove mathematical principles given suitable assumptions. But, of course, strong principles must be assumed concerning the existence of appropriate ordinal individuals. Instead of postulating strong existence principles *ab initio*, we may proceed more cautiously and assume just such principles as are needed for a given domain of mathematical theory. For the construction of the system of signed rational numbers, for example, a great deal more by way of existence is needed than for just the positive integers.

The various Boolean operations, for sums, products, and negations, are assumed available as well as an operation of unit-summation, where

> '(*x*1--*x*--)'

stands for the sum of all (cardinal or ordinal) entities *x* such that --*x*--. It is assumed that every *e* such that <Ord>*e* is a unit, in the sense that it is non-null and is a part of all its non-null parts. There are several alternative ways of handling all these matters technically that need not concern us here.

Given the simplest — i.e., unit — ordinal individuals, i.e., the *e*'s such that

> $(\mathrm{E}e_1)(\mathrm{E}e_2)e$ Ord e_1, e_2,

Boolean sums, products, and so on, of the term are also admitted. An *n*-adic relation in general, then, is any Boolean sum of unit *n*-adic ordinal individuals. These latter are any *e*'s such that

$$e \text{ Ord } x_1, \ldots, x_n,$$

where each x_i is a non-ordinal individual.

What, now, are we to say about the *agent* of an *e* such that <Ord>*e*? Are we to assume that all such acts are available somehow *sub specie aeternitatis* or are we explicitly to assume them to be performed by a human agent? or by a computer? or by an AI (artificial intelligence)? This too is a matter that may be left open for the moment, but very different conceptions of mathematics emerge depending upon how it is handled. The *sub specie aeternitatis* view is essentially that of classical mathematical "realism," with all its entities spread out before us in their pristine Fregean grandeur. If we relativize the acts to a specific person, say, to L.E.J. Brouwer himself, we have the means for building up an intuitionistic mathematics, depending upon just what <Ord>'s are taken as acceptable. And so on, for other views.

Let us take another step and treat the part-whole relation itself in terms of an event-descriptive predicate. Let

'<P>*e*'

be an act of taking something to be a part of something. Then

'*x* P *y*' is definable as '(E*e*)(<P>*e* . Act *e* . *e* $\text{On}_{\text{Object}}$ *x* . *e* WithRespectTo *y*)',

so that *x* is a part of *y* just where there is some act of taking it to be such.[5] In this way mereology itself takes on an act-theoretic form. Of course, the original entities *x*, *y*, and so on, are still "concrete" entities *in natura*, but that one is or is not a part of another is not a "fact," so to speak, but the result of a human (or other) act of some kind.

All of the Boolean notions can now be handled similarly, so that '<Sum>*e*' may be defined to express that *e* is an act of summing two entities, '<Prod>*e*', an act of forming the product, and so on. Boolean sums, products, and so on, are thus no longer assumed to exist independently of such acts, but only as the result of them.

Governing 'P' and the Boolean notations, the very general laws previously assumed to hold concerning them need no longer obtain. For example, we need no longer require that given any *x* and *y*, there is Boolean sum

$(x \cup y)$ of x and y. Instead only such sums exist as result from e's such that <Sum>e. And so on. And similarly for principles governing 'P' and the '1' of unit summation.

We have, then, two methods here, depending upon whether we take '<Ord>' and 'P' as the primitives, or '<Ord>' and '<P>'. The first might well be thought preferable to the latter, the notion of '<Ord>' seeming perhaps more natural than that of '<P>'.

Let us recall, from XVIII above, J.A. Wheeler's conception of Era III physics, "meaning physics," so called.[6] How, now, is Era III physics to differ from its predecessors? By bringing in fundamentally acts of observation, or the like, to provide somehow for the factor of "meaning." "Existence," according to Wheeler, "is the closed circuit of meaning.... Physics supplies to the machinery for constructing meaning, and meaning supplies the machinery for constructing physics. The first half of the "meaning circuit" is familiar in the main. Physics gives sight and sound and all the other agencies of communication. It also gives the biology to make the communicators! And out of communication comes meaning. The other half, the "return portion" ... is stranger. The demand for meaning we are to understand as the origin of probability and phase [in quantum mechanics], and phase we are to learn to consider as the definer of all fields, of all geometry, of all spacetime — and, in the end, therefore, of all particles and of all physics. Physics, on this view, is the child of meaning even as meaning is the child of physics."

'Meaning' here is not to be construed in a vague Frege-Carnap — or Montaguean, for that matter — way, but in terms of physically specifiable acts of observation, the full structure of which remains to be characterized. Here is surely a salient point that physicists have been neglecting all along. The stage seems to have been reached when deep philosophical — more specifically, semiotical — analysis of the full vocabulary of physics becomes imperative. And this analysis is no longer to be carried out in the amateurish way that is customary among philosophers, but only on the basis of a deep mastery of the details of physics in their most recent setting. A full semiotical and event-logical characterization of every detail of the physical language is now needed.

Further, meaning for Wheeler comes out of communication, which in turn presupposes communicators. A human agent seems thus needed, the physicist doing physics, in a way similar to that in which the "creative subjcet" is needed in some renditions of intuitionistic mathematics. Further, these doings are carried out at specific times and the results are usually explicitly written down in some fashion on tablets, or with pen and ink, and so on, in terms of physical inscriptions. No inscriptions, no written mathematics or physics, and thus no communication of such — except perhaps with extreme difficulty.

"No problem about "physics as meaning,"" Wheeler concludes, "strikes one as more immediate than time, and for no concept more than time are the implications more revolutionary. What about all those billions of years in the early universe when there were no communicators around to establish meaning? Were not those galaxies out there as real then as they are now? And equally real the galaxies which have not come into view today, but will swim up over the horizon tomorrow?" And again: "Essential in making headway into Era III physics is a reassessment of "time." Time is not a Heaven-sent concept. It is no luminous overarching sky supplied from outside for the benefit of physics. It is a human invention. The concept of "elasticity" is no more and no less a human invention. Elasticity fails, has no application, in the space between the electron and the atomic nucleus. Time likewise fails: fails at the Planck scale of distances, fails in the initial phase of the big bang, and fails in gravitational crunch. Both "time" and "elasticity" are not primordial entities in the description of nature, but secondary, approximate and derived (elasticity) or *yet to be derived* (time) from more primitive considerations. To supply that derivation is a central task — and opportunity — for Era III "meaning physics.""

If mathematics is developed in the fashion sketched above, in terms of '<Ord>' and 'P', or, alternatively, of, '<Ord>' and '<P>', as the basic primitives, the result is an Era III mathematics. Era I mathematics is the long history from the ancients up to about a century ago, and Era II is the period of the emergence and development of axiomatic set-theory and its applications. Era III no doubt has its beginnings in the work of Kant and Brouwer, although it is only quite recently that it has been given a formulation in terms of the "creative subject" — and, still more recently, seen that it could be formulated in terms of his or her explicit acts.

From these remarks, we see that mathematics and physics are much more closely related than is usually supposed. They have a common subject-matter, a common ontology, the physical objects. One salient difference emerges, however. Physics must make use of a multitude of primitive predicates applicable to the physical objects, whereas in mathematics no such predicates are needed. No, the only ones needed for the latter, as conceived here, are '<Ord>' and 'P' or '<Ord>' and '<P>'. Physics, however, needs these also, to provide its mathematical underpinning, as well as further ones, to provide for both the "meaning" and physical content.

What, now, are we to say about the existence of physico-mathematical objects, notions, and laws prior to the big bang and after gravitational collapse? Well, the very use of 'prior to' and 'after' here are undefined, and no definitions would seem forthcoming without a theory of time. There are at least two, quite distinct, ways of viewing the matter. We might regard the objects, notions, and laws as beginning their long evolution with the big bang and ending in the gravitational crunch — and that is that. Or, alternatively, we can regard the event-descriptive notions as "Heaven-sent concepts," in Wheeler's very phase, having a non-physical existence and content of some sort. Of course, there are no physical objects for them to be applicable to — unless physical objects themselves be handled in terms of a still further kind of event-descriptive predicate of a non-physical kind. Such a view is not so *outré* as it might appear. C.S. Peirce, for example, has espoused something similar — matter as mind "hidebound with habit" — and objective idealists throughout the centuries have spoken, in varying ways, of mind as "embodied" in material objects. Although this second view transcendends the limitations of what ordinarily passes as either physics or mathematics, the fundamentals of the Era III conception of them would seem to be in essential harmony with it.

Appendix: a thumbnail sketch of ordinal mereology

"Sets are central," Meyer and Routley have told us, "to extensionalist formal ontology. It was not always so ... for many were the reductionist projects which sought to construct the whole World out of particulars of some kind or other. It is generally agreed, however, that these projects did not come off — at least they did not come off well enough to provide for mathematics, which everyone agrees should be provided for." Presumably the authors are referring to modern forms of nominalism that renounce sets and employ only extensionalist ontologies. Later they ask "Can the extensionalist do without sets? This is, as they say, the heroic course." Some of us have resolutely been following the heroic course over the years, and moreover, far enough to provide for some portions of mathematics, portions comparable to those provided by Brouwerian intuitionism in its modern forms. An example is the author's "A Homogeneous System [H] for Formal Logic" and the two papers that followed (*J.S.L.*, vols. 8, 13, & 14). In the system H, it will be recalled, no use is made of sets, only of individuals being admitted as values for variables. Of course abundant use is made of *virtual* classes and relations, but this is a very different matter. The system H consists essentially of a mereology or calculus of individuals based on a part-whole relation P and augmented with a theory of ancestral quantification. This latter is now dropped, in the present paper, in favor of a triadic relation Ord providing for the notion of an *ordinal* individual. The inclusion of this latter seems a very natural extension of mereology. 'Ord' is taken as a primitive significant in contexts of the form '*x* Ord *y*,*z*', expressing that *x* consists of the individuals *y* and *z* *taken in this order* — in other words, that *x* is an ordinal individual, a kind of non-communicative quasi-mereological compound of *y* and *z*. The calculus of individuals is augmented with the theory of such ordinals, so that the ordinary Boolean sums, products, etc., of them may also be formed. In this way relations of various degrees between or among individuals may be accommodated, such relations being ordinary Boolean sums of ordinal individuals. And similarly for relations between relations. In this paper it is shown how such a framework can be made to provide a foundation for numerical mathematics up to and including the complex number-system. If these considerations — or something closely akin — are sound, we would see to be on the road to gaining the simplest, most straightforward, and ontologically most scrupulous foundation for mathematics yet put forward.

In addition to 'P' and 'Ord', let expressions of the form '$(x1{-}{-}{-}x{-}{-}{-})$' be primitive as terms for the Boolean sum of all unit or atomic individuals x such that $--x--$. Similarly, $\{x\ni --x--\}$ is the virtual class of all individuals x such that $--x--$; and likewise for virtual relations. The following definitions and principles (with omissions) are then more or less self-explanatory.

Pr1. $\vdash (x)x \mathrm{P}\, x$,

Pr2. $\vdash (x)(y)(z)((x \mathrm{P}\, y \,.\, y \mathrm{P}\, z) \supset x \mathrm{P}\, z)$,

D1. '$x = y$' abbreviates '$(x \mathrm{P}\, y \,.\, y \mathrm{P}\, x)$',

Pr3a. $\vdash (x)(y)(z)(w)((x \,\mathrm{Ord}\, y,z) \,.\, w \,\mathrm{Ord}\, y,\, z \supset x \mathrm{P}\, w)$,

Pr3b. $\vdash (x)(y)(z)(w)((x \,\mathrm{Ord}\, y,z) \,.\, w \,\mathrm{Ord}\, y,\, z \supset y \mathrm{P}\, w)$,

Pr3c. $\vdash (x)(y)(z)(w)((x \,\mathrm{Ord}\, y,z) \,.\, w \,\mathrm{Ord}\, y,\, z \supset z \mathrm{P}\, w)$,

Pr4. $\vdash (x)(y)(x = y \supset (z)(u)((x \,\mathrm{Ord}\, z,\, y \equiv y \,\mathrm{Ord}\, z,u) \,.\, (z \,\mathrm{Ord}\, x,u \equiv z \,\mathrm{Ord}\, y,u) \,.\, (z \,\mathrm{Ord}\, u,x \equiv z \,\mathrm{Ord}\, u,y)))$,

Pr5. $\vdash (x)(y)(z)(w)(u)(v)((x \,\mathrm{Ord}\, y,z \,.\, w \,\mathrm{Ord}\, u,v) \supset (x = w \equiv ((y = u \,.\, z = v) \vee (\mathrm{E}t)((y \,\mathrm{Ord}\, u,t \,.\, v \,\mathrm{Ord}\, t,z) \vee (z \,\mathrm{Ord}\, t,v \,.\, u \,\mathrm{Ord}\, y,t)))))$,

D2. '$(x \cup y)$' abbreviates '$(z1(z \mathrm{P}\, x \vee z \mathrm{P}\, y))$',

D3. '$(x \cap y)$' abbreviates '$(z1(z \mathrm{P}\, x \,.\, z \mathrm{P}\, y))$',

D4. '$-x$' abbreviates '$(y1 \sim y \mathrm{P}\, x)$',

D5. 'W' abbreviates '$(x1x \mathrm{P}\, x)$',

D6. 'N' abbreviates '-W',

D7a. 'OrdInd x' abbreviates '$(\mathrm{E}y)(\mathrm{E}z)x \,\mathrm{Ord}\, y,z$',

D7b. 'CardInd x' abbreviates '$x \mathrm{P}\, (y1 \sim \mathrm{OrdInd}\, y)$',

D8. 'Unit x' abbreviates '$(\sim x \mathrm{P}\, \mathrm{N} \,.\, (y)((\sim y \mathrm{P}\, \mathrm{N} \,.\, y \mathrm{P}\, x) \supset x \mathrm{P}\, y))$',

Pr6a. $\vdash (x)(\mathrm{CardInd}\, x \supset (x \mathrm{P}\, (y1(\mathrm{CardInd}\, y \,.\, --y--)) \equiv (y)((\mathrm{Unit}\, y \,.\, \mathrm{CardInd}\, y \,.\, y \mathrm{P}\, x) \supset --y--)))$,

Pr6b. $\vdash (x)(\mathrm{CardInd}\, x \supset ((y1(\mathrm{CardInd}\, y \,.\, --y--)) \mathrm{P}\, x \equiv (y)((\mathrm{Unit}\, y \,.\, \mathrm{CardInd}\, y \,.\, --y--) \supset y \mathrm{P}\, x)))$,

Pr7. $\vdash (\mathrm{E}x)(\mathrm{CardInd}\, x \,.\, \mathrm{Unit}\, x)$,

D9a. '$x \,\mathrm{Ord}\, y,z,w$' abbreviates '$(\mathrm{E}u)((u \,\mathrm{Ord}\, y,z \,.\, x \,\mathrm{Ord}\, u,w) \vee (u \,\mathrm{Ord}\, z,w \,.\, x \,\mathrm{Ord}\, y,u))$', etc.,

D10a. '(x ord y) P z' abbreviates '(Ew)(w Ord x,y . w P z)',

etc.,

Pr8. ⊢ (x)(y)((CardInd x . CardInd y) ⊃ (Ez)z Ord x,y),

Pr9. ⊢ (x)(y)...(z)((CardInd x . CardInd y . ... . CardInd z) ⊃ (Ew)w Ord x,y,...z), for any number of quantifiers ≧ 3,

Pr10. ⊢ (x)(y)(z)(x Ord y,z ⊃ (∼ x P y . ∼ x P z)),

Pr11. ⊢ (x)(y)(z)(x Ord y,z ⊃ ∼ (∼ y P N . y P x) . ∼ (∼ z P N . z P x))),

Pr12. ⊢ (x)(y)((OrdInd x . ∼ y P N . y P x) ⊃ x P y),

D11. 'x Suc y' abbreviates '(CardInd x . CardInd y . (Ez)(CardInd z . Unit z . (y ∩ z) P N . x = (y ∪ z)))',

Pr13. ⊢ (x)(CardInd x ⊃ (Ey)y Suc x)),

D12a. 'DyRel(Card) x_2' abbreviates '(∼ x P N . (y)((Unit y . y P x) ⊃ (Ez)(Ew)(y Ord z,w . CardInd z . CardInd w)))',

D12b. 'TriRel(Card) x_3' abbreviates '(∼ x P N . (y)((Unit y . y P x) ⊃ (Ez)(Ew)(Eu)(y Ord z,w,u . CardInd z . CardInd w . CardInd u)))',

etc.,

D13a. 'x z_2 y' abbreviates '(DyRel(Card) z_2 . (x ord y) P z_2)',

D13b. 'x w_3 y,z' abbreviates '(TriRel(Card) w_3 . (x ord y ord z) P w_3)',

etc.,

Pr14a. ⊢ (x_2)(DyRel(Card) x_2 ⊃ (x_2 P (z1(Ew)(Eu)(CardInd w . CardInd u . z Ord w,u . --w--u--)) ≡ (w)(u)(z)((CardInd w . CardInd u . z Ord w,u . z P x_2) ⊃ --w--u--))),

Pr14b. ⊢ (x_2)(DyRel(Card) x_2 ⊃ ((z1(Ew)(Eu)(CardInd w . CardInd u . z Ord w,u . --w--u--)) P x_2 ≡ (w)(u)(z)((CardInd w . CardInd u . z Ord w,u . --w--u--) ⊃ z P x_2))),

D15a. 'D'w_2' for '{x ∋ (Ey)x w_2 y}',

D15b. 'D'w_3' for '{x ∋ (Ey)(Ez)x w_3 y,z}',

etc.,

D16a. 'Ɑ'w_2' for '{y ∋ (Ex)x w_2 y}',

D16b. 'Ɑ'w_3' for '{yz ∋ (Ex)x w_3 y,z',

etc.,

D17. '(1→Cls)w_2' for '(x)(y)(z)((x w_2 y . z w_2 y) ⊃ x = z)',

D18. '(Cls→1)w_2' for '(x)(y)(z)((x w_2y . x w_2 z) ⊃ y = z)',

D19. '(1→1) w_2' for '((1→Cls)w_2 . (Cls→1)w_2)',

D20. '*x* Sm *y*' for '(CardInd *x* . CardInd *y* . (Ew_2)((1→1) w_2 . D'w_2 = {*z* ∋ (Unit *z* . CardInd *z* . *z* P *x*)} . D'w_2 = {*z* ∋ (Unit *z* . CardInd *z* . *z* P *y*)}))',

D21. 'Nc'*x*' abbreviates '{*y* ∋ *y* Sm *x*}',

D22a. '0' abbreviates 'Nc'N',

D22b. '1' abbreviates ' {*x* ∋ (E*y*)(Unit *y* . *x* Sm *y*)}',

etc.,

D23. 'u_2 Att *x*' abbreviates '(*x* u_2 *x* . (*z*)(*w*)(*z* u_2 *w* ⊃ *z* = *w*))',

D24. '*x* *Suc *y*' abbreviates '(u_2)(u_2 Att *y* . (*z*)(*w*)((u_2 Att *w* . *z* Suc *w*) ⊃ u_2 Att *z*)) ⊃ u_2 Att *x*)',

D25. 'Fin *x*' abbreviates '*x* *Suc N',

D26. 'FinNc'*x*' or '$\bar{\bar{x}}$' abbreviates '{*y* ∋ (Fin *x* . *y* Sm *x*)}',

D27. '($\bar{\bar{x}} + \bar{\bar{y}}$)' abbreviates '{*z* ∋ (E*w*)(E*u*)(*w* ε $\bar{\bar{x}}$. *u* ε $^{=}y$. (*w* ∩ *u*) P N . *z* = (*w* ∪ *u*))}',

D28. '($\bar{\bar{x}}$)--$\bar{\bar{x}}$--' abbreviates '(*x*)(Fin *x* ⊃ --*x*--)', where (as needed),

D29. '(E$\bar{\bar{x}}$)--$\bar{\bar{x}}$--' abbreviates '~ ($\bar{\bar{x}}$) ~ --$\bar{\bar{x}}$--',

D30. '(ι$\bar{\bar{x}}$. --$\bar{\bar{x}}$--)' abbreviates '{*y* ∋ (E$\bar{\bar{x}}$)(--$\bar{\bar{x}}$-- . ($\bar{\bar{z}}$)(--$\bar{\bar{z}}$-- ⊃ $\bar{\bar{z}} = \bar{\bar{x}}$) . *y* ε $\bar{\bar{x}}$)}', where (as needed),

D31. '$\bar{\bar{x}}, \bar{\bar{y}}, \bar{\bar{z}}$ Mult $\bar{\bar{x}}', \bar{\bar{y}}', \bar{\bar{z}}'$' abbreviates '(Fin *x* . Fin *y* . Fin *z* . Fin *x'* . Fin *y'* . Fin *z'* . $\bar{\bar{x}} = (\bar{\bar{x}}' + \bar{\bar{y}}')$. $\bar{\bar{y}} = \bar{\bar{y}}'$. $\bar{\bar{z}} = (\bar{\bar{z}}' + 1)$)',

D32. '$^{=}x, ^{=}y, ^{=}z$ *Mult $^{=}x', ^{=}y', ^{=}z'$' abbreviates '(w_3)((Fin *x* . Fin *y* . Fin *z* . Fin *x'* . Fin *y'* . Fin *z'* . *x'* w_3 *y'*, *z'* . (*t*)(*u*)(v)(*w'*)(*u'*)(*v'*)((*w'* w_3*u'*,*v'* . $^{=}t, ^{=}u, ^{=}v$ Mult $^{=}w', ^{=}u', ^{=}v'$) ⊃ *t* w_3 *u*,*v*)) ⊃ *x* w_3 *y*,*z*)',

D33. '($^{=}x \times {}^{=}y$)' abbreviates '(ι$^{=}z$. $^{=}z, ^{=}x, ^{=}y$ *Mult 0,$^{=}x$,0)'.

Pr1-Pr2 stipulate that P is a *totally reflexive* and *transitive* relation. *Pr3a-Pr5* are Principles of Identity and *Pr6a-Pr6b* are restricted Principles of Unit Summation. *Pr7-Pr9* are fundamental Principles of Existence. *Pr10-Pr12* are Principles of Ordination, and *Pr13* is a kind of Axiom of Infinity for cardinal individuals. *Pr14a-Pr14b* are Principles of Unit Summation for Ordinal Units. The definitions *D1-D6* are of a familiar kind, *D2-D6* being of the basic *Boolean* notions. *D7a* introduces the notion of an *ordinal individual*, as contrasted with that of a *cardinal* individual, in *D7b*. The completely general notion of being a *unit* individual in introduced in *D8*. In *D9a et seq.*, a notation for (unit) *triadic, tetradic*, and so on, *relational individuals* is provided, and in *D10a et seq.*, a notation for '(*x* ord *y*)', '((*x* ord *y*) ord *z*)', etc., as *terms* occuring either side of 'P'. *D11* introduces the *successor-relation* as between cardinal individuals. *D12a-D13b et seq.* introduce the notions of being a *dyadic* relation-individual between cardinals, a triadic one, and so on.

D15a et seq. provide for the notions of being the *domain* of a dyadic, triadic, and so on, relation-individual, and *D16a et seq.*, those for the *converse-* or *counterdomain* of relation-individuals. The converse domains for triadic, and so on, relation-individuals are construed, it should be noted, relationally. Note also that domains and counterdomains are provided for in terms of *virtual* classes and relations. *D17-D19* give us the notions of a *one-many, many-one*, and *one-one* dyadic relation-individual or "correspondence," preparatory to the definition, *D20*, of the notion of being *similar to*, or *equinumerous with*, as between cardinal individuals. *D21* provides the notion of being a *cardinal number* of a cardinal individual, and *D22a et seq.* provide specifically for 0, 1, etc.

The general notion of being an *attributive relation*, as applying to cardinal individuals, is defined in *D23*, and that of the *ancestral* of the successor-relation, in *D24*. The notion of being a *finite* cardinal individual is provided by *D25*, and a notation for *finite cardinal numbers*, in *D26*. *D27* introduces a notation for the *sum of two finite cardinal numbers*. *D28-D30* introduce the *quantifiers* and *definite descriptions* for finite cardinal numbers. Finally, *D31-D32* introduce a *product* of finite cardinal numbers in terms of the ancestral of a certain hexadic relation Mult.

After all of this, we can move on to the positive rationals, the signed rationals, the reals, and to the complex numbers in more or less familiar ways, on the basis, of course, of such further existence and other principles as are needed. Various areas of applied logic and mathematics may then be provided by the addition of suitable non-logical constants and principles.

This Appendix is a slightly expanded version of the abstract of the paper "On Mereology and the Heroic Course," presented by invitation at the VIIth International Congress for Logic, Methodology, and Philosophy of Science, at Salzburg in July, 1983, and published in the *Abstracts* thereof (J. Hutteger OHG, Salzburg: 1983), pp. 178-180. It is reproduced here by kind permission.

Notes to Chapter One

1. The phrase is borrowed from Edward S. Klima, "Negation in English," in *The Structure of Language*, ed. by J.A. Fodor and J.J. Katz (Prentice-Hall, Incl., Englewood Cliffs, N.J.: 1964), pp. 246-323.

2. On this latter, see the forthcoming *Handbook of Philosophical Logic* (D. Reidel Publishing Co., Dordrecht: to appear), a work probably largely out-of-date even prior to publication, as such "surveys" tend to be.

3. Recall Chapter 17, "'And'", in *Pragmatics, Truth, and Language* (D. Reidel Publishing Co., Dordrecht: 1979).

4. Recall esp. Chapter VII, "Of Time and the Null Individual," in *Pragmatics, Truth, and Language.*

5. Recall esp. *Belief, Existence, and Meaning* (New York University Press, New York: 1969), Chapter VI, "The Philosophic Import of Virtual Classes," and *Semiotics and Linguistic Structure* (State University of New York Press, Albany: 1978), Chapters I and II.

6. On the theory of relations in general, recall Chapter XIII, "Some Phases in the History of the Logic of Relations," in *Metaphysical Foundations: Mereology and Metalogic* (Philosophia Verlag, München: to appear). On Quine's corners, see his *Mathematical Logic* any edition.

7. For more on mereology, recall *Metaphysical Foundations: Mereology and Metalogic*, and see several of the papers below.

8. (Wissenschaftliche Buchgesellschaft, Darmstadt: 1965).

9. See esp. I.M. Bochénski, *A History of Formal Logic* (Notre Dame University Press, Notre Dame: 1961).

10. See esp. A. Tarski, *Logic, Semantics, Metamathematics* (Clarendon Press, Oxford: 1956), pp. 165 ff., and the author's *Truth and Denotation* (University of Chicago Press, Chicago: 1958), Chapter III, and *Semiotics and Linguistic Structure*, Chapter V.

11. Recall *Truth and Denotation*, Chapters IV and V, and *Semiotics and Linguistic Structure*, Chapter VI.

12. As in Tarski, *op. cit.*, etc. See also Tarski's popular paper in *Philosophy and Phenomenological Research* IV (1944): 341-376.

13. Recall Chapter I of *Pragmatics, Truth, and Language*.

14. Cf. Chapter XVI, "On the Philosophical Challenge of Intuitionistic Logic," in *Mind, Morality, Meaning, and Method* (State University of New York Press, Albany: 1983).

15. Many such relations were apparently first explicitly called attention to in pragmatical terms by Frege. Recall Chapter VII, "Frege's Pragmatic Concerns" In *Peirce's Logic of Relations and Other Studies* (Foris Publications, Dordrecht: 1980).

16. See Frege's famous "Über Sinn und Bedeutung", the second paragraph, and *Begriffsschrift*, §8, in *Translations from the Philosophical Writings of Gottlob Frege*, ed. by P. Geach and M. Black (Basil Blackwell, Oxford: 1952), and the author's *Mind, Modality, Meaning, and Method*, Chapter III and IV, "On Being quā Being, Protometaphysics, and Epistemic Modality" and "On Actuality and Objective Modality."

17. Note the kinship here with the *objective quasi-intensions* of *Toward a Systematic Pragmatics* (North-Holland Publishing Co., Amsterdam: 1959). See also III below.

18. Recall H. Reichenbach, *Elements of Symbolic Logic* (The Macmillan Co., New York: 1947), pp. 301 ff., and R. Montague, *Formal Philosophy* (Yale University Press, New Haven: 1974), esp. p. 191.

19. Recall, e.g., Chapter XV, "On Minds and Their Brains: Sir John Eccles," in *Mind, Modality, Meaning, and Method*.

20. Recall Chapter XIII, "The One and the Many: Logic_1 and Logics_2," in *Mind, Modality, Meaning, and Method*.

21. As in the *Logische Untersuchungen* (Max Niemeyer, Halle). Recall Chapter XI, "A Prolegomena to Husserl's *Prolegomena*," in *Mind, Modality, Meaning, and Method*, and Chapter XVIII, "On the Concept of Truth in Husserl's *Logische Untersuchungen," in Metaphysical Foundations: Mereology and Metalogic*.

22. On prepositional relations, recall *Semiotics and Linguistic Structure*, Chapters X, XI, and XII.

23. See, e.g., Otto Jesperson, *Negation in English and Other Languages* (Copenhagen: 1917).

24. See Zellig Harris, *A Grammar of English on Mathematical Principles* (John Wiley and Sons, New York: 1982).

25. See R. Carnap, *Meaning and Necessity* (University of Chicago Press: 1956), second edition, pp. 222 ff.

26. Cf. *Truth and Denotation*, Chapters I and XIII.

27. *Op.cit.*, p. 125.

Notes to Chapter Two

1. William Chauncey Fowler, *The English Language in Its Elements and Forms* (Harper and Brothers, New York: 1870), p. 238.

2. Recall especially *Events, Reference, and Logical Form* (The Catholic University of America Press, Washington: 1978); *Semiotics and Linguistic Structure; Pragmatics, Truth, and Language*; and *Logico-Linguistic Papers* (Foris Publications, Dordrecht: 1981).

3. In *Language in Society* 12 (1983): 286-287.

4. See Richard Montague, *op.cit.*

5. In *Pragmatics, Truth, and Language*.

6. First published in 1786.

7. Recall "The Philosophical Import of Virtual Classes" in *Belief, Existence, and Meaning*, and Chapters I and II of *Semiotics and Linguistic Structure.*

8. Recall especially Chapters XI of *Truth and Denotation* and V of *Semiotics and Linguistic Structure*.

9. Recall Chapters IV and V of *Truth and Denotation* and VI *Semiotics and Linguistic Structure*.

10. Recall especially Chapters III and IV of *Mind, Modality, Meaning, and Method*, and VII of *Logico-Linguistic Papers*.

11. Recall Chapters XI and XII of *Semiotics and Linguistic Structure*.

12. See R. Carnap, *Introduction to Semantics* (Harvard University Press, Cambridge: 1942), p. 90.

13. Recall "On Mereology and the Heroic Course," in *Metaphysical Foundations: Mereology and Metalogic,* and see V and the Appendix below.

14. (Mouton and Co., the Hague: 1973), p. 7.

Notes to Chapter Three

1. *Encyclopedia of Unified Science*, Vol. I, No. 3 (University of Chicago Press, Chicago: 1939).

2. (Harcourt, Brace and Co., New York: 1937).

3. (Harvard University Press, Cambridge: 1942).

4. *Philosophical Studies*, 6 (1955): 89-91, and reprinted in *Meaning and Necessity*, 2nd ed. (University of Chicago Press, Chicago: 1956), pp. 248-250. (The first edition was published in 1947.)

5. *The Journal of Symbolic Logic*, 12 (1947): 105-122.

6. *Ibid.*, 16 (1951): 191-203.

7. (University of Chicago Press, Chicago: 1958 and 1976).

8. See, e.g., William Cooper, *Logico-Linguistics* (D. Reidel Pub. Co., Dordrecht: 1979), and *Semiotics and Linguistic Structure*, part B.

9. (*Studies in Logic and the Foundations of Mathematics*, North-Holland Publishing Co., Amsterdam: 1959).

10. A. Tarski, *Logic, Semantics, Metamathematics*, Chapter VIII.

11. W.V. Quine, *Mathematical Logic*, any edition, Chapter VII.

12. Recall *Truth and Denotation*, Chapters VIII and IX, and see VI below.

13. Recall "Über Sinn und Bedeutung," second paragraph, and *Begriffsschrift*, §8.

14. On Under, recall especially *Mind, Modality, Meaning, and Method*, Chapters III and IV.

15. For further details, recall *Events, Reference, and Logical Form; Semiotics and Linguistic Structure; Pragmatics, Truth, and Language*; and *Logico-Linguistic Papers*. In the forego-

ing the notation $\ulcorner \hat{x}^1 a \urcorner$ (or $\ulcorner \hat{e} a \urcorner$) — for the real classes of PM^T — is used rather than the $\ulcorner x^1 \ni a \urcorner$ of *TSP* because $\ulcorner \{x \ni a\} \urcorner$ (or $\ulcorner \{e \ni a\} \urcorner$) became the preferred one for virtual classes, in the books just mentioned and in I and II above. The '*e*'s are event-variables and of lowest type.

Notes to Chapter Four

1. P.T. Geach, "Comparatives," *Philosophia* 13 (1983): 235-246.
2. See esp. his *Logical Foundations of Probability*, pp. 8 ff.
3. Recall *Events, Reference, and Logical Form, Semiotics and Linguistic Structure*, and so on.
4. On transformations, about which nothing is said here, recall "On Some Protolinguistic Transformations" in *Pragmatics, Truth, and Language*, and "On Logico-Linguistics: Structure and Transformation" in *Logico-Linguistic Papers.*
5. On the logic of 'very' recall *Semiotics and Linguistic Structure*, p. 159.
6. Recall "On Metaphor and Logical Form," in *Logico-Linguistic Papers.*
7. Recall "On the Logic of Fictive Locutions", also in *Logico-Linguistic Papers.*
8. Recall *Semiotics and Linguistic Structure*, pp. 73, 176, and 200 ff., and see XVIII below.
9. Recall "On the Logical Structure of the Ontological Argument," in *Whitehead's Categoreal Scheme and Other Papers* (Martinus Nijhoff, the Hague: 1974).

Notes to Chapter Five

1. See "A Homogeneous System for Formal Logic," *The Journal of Symbolic Logic*, Vol. 8: 1-23 and "On Mereology and the Heroic Course," in *Metaphysical Foundations: Mereology and Metalogic*. The system of this latter paper is referred to as OM, an *ordinal* mereology, rather than as EM. For a thumbnail sketch of it, see the Appendix below.
2. Czeslaw Lejewski, "On Lesniewski's Ontology," *Ratio*, 1 (1958): 150-176. Cf. also John T. Kearns, "The Contribution of Lesniewski," *Notre Dame Journal of Formal Logic* VIII(1-2): 61-93.
3. J.H. Woodger, *Biology and Language* (Cambridge University Press, Cambridge: 1952), p. 17, and "Science without Properties," *British Journal for the Philosophy of Science* II (1952): p. 196.
4. Recall *90 of *Principia Mathematica*.
5. In "On the Algebra of Relations and Their Affiliates," in *Mind, Modality, Meaning, and Method*. See also XIII below.
6. Strictly, it should be noted, expressions '$(x1\text{--}x\text{--})$' for unit sums are assumed now to be introduced by definition rather than primitively.

Notes to Chapter Six

1. C. Lejewski, *op.cit.*.
2. Recall *Truth and Denotation*, Chapters VIII and IX.
3. Recall Chapter XVI of *Metaphycical Foundations: Mereology and Metalogic*.
4. Recall *Truth and Denotation*, Chapter III.
5. These are of course spelled out in primitive terms, with omission of the parentheses not needed due to the associativity of '$\cap$'.
6. Recall *Belief, Existence, and Meaning*, Chapter II, "The Elimination of Intensions *Sui Generis*."
7. Z. Stachniak, *Introduction to Model Theory for Lesniewski's Ontology* (Acta Universitatis Wratislaviensis, No. 586, Prace Filozoficzne XXXI, Logika 9: Wroclaw (1981)).

Notes to Chapter Seven

1. *L'enseignements mathématiques* 34 (1935): 52-69.
2. Cf. "Some Remarks on the Language of Mathematical Intuitionism" in *Events, Reference, and Logical Form*, and "On the Philosophical Challenge of Intuitionistic Logic" in *Mind, Modality, Meaning, and Method*.
3. Cf. "On *Eidos*, Instance, and Aspect" and "Some Phases in the History of the Logic of Relations," in *Metaphysical Foundations: Mereology and Metalogic*.
4. In his *Plato, the Man and His Work* (The Dial Press, New York: 1936), p. 50.
5. In "On Mereology and the Heroic Couse," in *Metaphysical Foundations: Mereology and Metalogic*, and Chapter V above.
6. On virtuality see "The Philosophic Import of Virtual Classes" in *Belief, Existence, and Meaning*, and *Semiotics and Linguistic Structure*, Chapters I and II.
7. Cf. also C.S. Peirce, *Collected Papers*, 3.492, and A. Tarski, *Logic, Semantics, Metamathematics*, p. 334.
8. The '℩' here is the familiar inverted iota and helps to symbolize Russellian descriptions.
9. Not to be confused with the great and famous theorem due to Dr. Brook Taylor (1715) concerning the convergence of a power series as always at least so large as only just to exclude from the interior of the circle of convergence the nearest singularity of the function represented by the series.
10. In "Forms and Numbers: A Study in Platonic Metaphysics," *Mind* XXXVI (1927): 12-33.
11. (Almquist & Wiksell, Stockholm: 1955), pp. 65 ff.

Notes to Chapter Eight

1. (W.W. Norton and Co., New York: 1938 but first published in 1903), pp. 23 f.
2. Recall "On the Algebra of Relations and Their Affiliates," in *Mind, Modality, Meaning, and Method*, and other relevant literature.
3. C.S. Peirce, "On the Algebra of Logic," *Americal Journal of Mathematics* 3 (1880): 15-57, and "On the Algebra of Logic: A Contribution to the Philosophy of Notation," *Ibid.*, 7 (1885): 180-202. Both of these reappear in the *Collected Papers*, Vol. III, Papers VI and XIII.
4. Recall "The Elimination of Intensions *Sui Generis*," in *Belief, Existence, and Meaning.*
5. See A. Tarski, *Logic, Semantics, Metamathematics*, Chapter XI.
6. N. Wiener, "A Simplification in the Logic of Relations," *Proceedings of the Cambridge Philosophical Society* 17 (1912-1914): 387-390, and C. Kuratowski, "Sur la Notion de l'Ordre dans la Théorie des Ensembles," *Fundamenta Mathematicae* 2 (1921): 161-171. The methods of these two authors are not the same.
7. See their *Grundlagen der Mathematik*, vol. 1, §5.
8. On the hoary distinction between the abstract and the concrete, recall some of the comments in "On Metatheory-Formulation and the World in the Making," in *Metaphysical Foundations: Mereology and Metalogic*. On sets see also X below.
9. Recall "De Morgan and the Logic of Relations," in *Peirce's Logic of Relations and Other Studies.*
10. A definition not so very simple, be it said.
11. *Isis*, XIX, 1: 285-297.
12. "The Logic of Relatives," in *Studies in Logic by Members of the Johns Hopkins University* (Little, Brown & Co., Boston: 1883), reprinted as XII in Vol. III of the *Collected Papers.*

Notes to Chapter Nine

1. Nelson Goodman, "A World of Individuals," in *The Problem of Universals* (University of Notre Dame Press, Notre Dame: 1956), and *The Structure of Appearance* (Harvard University Press, Cambridge: 1951). See also Goodman and Quine, "Steps toward a Constructive Nominalism," *The Journal of Symbolic Logic*, 12 (1947): 105-122.
2. *The Philosophical Review*, 62 (1953): 108-116.
3. As in Chapter XX of *Metaphysical Foundations: Mereology and Metalogic.*
4. Recall the author's "The Principle of Nominalism," *Philosophical Studies*, XIV (1963): 33-37, and "Mathematical Nominalism" in *Pragmatics, Truth, and Language.*
5. (D. Reidel Publishing Co., Dordrecht: 1970), pp. 81-99.
6. *Foundations of Concept Formation in Empirical Science* (*International Encyclopedia of Unified Science*, Vol. II, No. 7, The University of Chicago Press, Chicago: 1952), p. 21.

Notes to Chapter Ten

1. Hao Wang, *From Mathematics to Philosophy* (Humanities Press, New York: 1974).
2. Recall "Set Theory and Royce's Modes of Action" in *Peirce's Logic of Relations and Other Studies*, and "On Meaning, Protomathematics, and the Philosophy of Nature" in *Primordiality, Science, and Value*.
3. Kurt Gödel, "What Is Cantor's Continuum Problem?" in *Philosophy of Mathematics, Selected Readings* (Prentice-Hall, Inc., Englewood Cliffs, N.J.: 1964), pp. 258-273 (p. 287), reprinted from the *American Mathematical Monthly* 54 (1947): 515-525.
4. In *The Philosophy of Bertrand Russell* (*The Library of Living Philosophers*, Vol. V; Northwestern University, Evanston and Chicago: 1944), pp. 123-153.
5. Recall "A Prolegomenon to Husserl's *Prolegomena*" in *Mind, Modality, Meaning, and Method*, and "On the Concept of Truth in Husserl's *Logische Untersuchungen*" in *Metaphysical Foundations: Mereology and Metalogic*.
6. Recall "Toward an Objectual Dynamics of Logical Systems," in *Logico-Linguistic Papers*.
7. *Royce's Logical Essays* (Wm. C. Brown Company, Dubuque, Iowa: 1951), p. 350.
8. Recall "On the Philosophical Import of Virtual Classes," in *Belief, Existence, and Meaning*.
9. In the *Review of Metaphysics* 35 (1981): p. 396.
10. (John Wiley and Sons, New York: 1982).

Notes to Chapter Eleven

1. "Scientific Causal Talk: a Reply to Martin," *Theory and Decision* 13 (1981): 363-379.
2. "On the Language of Causal Talk," in *Pragmatics, Truth, and Language*, pp. 192-205.
3. For a full characterization, recall the material of *Semiotics and Linguistic Structure; Events, Reference, and Logical Form; Primordiality, Science, and Value*; and *Logico-Linguistic Papers*.
4. As in I above and "On Mereology and the Heroic Course," in *Metaphysical Foundations: Mereology and Metalogic*.
5. "The Plurality of Science," in *PSA 1978*, Vol. 2, ed. by P. Asquith and I. Hacking (Philosophy of Science Association, East Lansing: 1981).
6. In the Introduction to R. Carnap, *The Unity of Science* (Kegan Paul, Trench, Trubner & Co., London: 1934), p. 13.
7. "Procedural Semantics," in *Language, Logic, and Philosophy*, Proc. 4th Inter. Wittgenstein Symposium (Hölder-Pichler-Tempski, Vienna: 1980), pp. 27-35.
8. As, for example, in *Truth and Denotation*.
9. See XII below.

10. Recall *Semiotics and Linguistic Structure*, the three chapters on prepositional relations and prepositional protolinguistics.

11. Recall *Semiotics and Linguistic Structure*, p. 246.

Notes to Chapter Twelve

1. Philip J. Davis and Reuben Hersh, *The Mathematical Experience* (Birkhäuser, Boston, Basel, Stuttgart: 1981).

2. Recall *Truth and Denotation*, pp. 167, 170, 172 and *Semiotics and Linguistic Structure*, pp. 107 ff.

3. Recall *Mind, Modality, Meaning, and Method*, Chapter XI, for a discussion of an example from Husserl.

4. (Cambridge University Press, Cambridge: 1976).

5. Recall *Mind, Modality, Meaning, and Method*, Chapter XIII, "The One and the Many: Logic_1 and Logics_2."

6. G. Poyla, *How to Solve It* (Princeton University Press, Princeton: 1945).

7. Recall Chapter XVI, "On the Philosophical Challenge of Intuitionistic Logic," in *Mind, Modality, Meaning, and Method.*

8. See P. Martin-Löf, "Constructive Mathematics and Computer Programming," in *Logic, Methodology, and Philosophy of Science VI* (*Studies in Logic and the Foundations of Mathematics*, North-Holland Publishing Co., Amsterdam, New York, Oxford and PWN-Polish Scientific Publishers, Warszawa: 1982), pp. 151-175.

Notes to Chapter Thirteen

1. "On the Algebra of Relations and their Affiliates," in *Mind, Modality, Meaning, and Method.*

2. Recall "The Philosophic Import of Virtual Classes" in *Belief, Existence, and Meaning*, and Chapters I and II of *Semiotics and Linguistic Structure.*

3. *Grundlagen der Mathematik* (Springer Verlag, Berlin: 1934), pp. 164 ff.

4. Cf. A. Church, *Introduction to Mathematical Logic* (Princeton University Press, Princeton: 1956), §42.

5. Cf. G. Frege, *On the Foundations of Geometry and Formal Theories of Arithmetic* (Yale University Press, New Haven: 1971), p. 112 and *passim.*

6. Recall *Truth and Denotation*, Chapter III and VIII, and *Semiotics and Linguistic Structure*, Chapter V.

7. *Collected Papers*, 3.45-3.149. See also "Of Lovers, Servants, and Benefactors", in Peirce's *Logic of Relations and Other Studies*, p. 41.

8. *Collected Papers*, 3.328-3.403A.

9. See especially Haskell B. Curry and Robert Feys, *Combinatory Logic I* (North-Holland Publishing Co., Amsterdam: 1958), and Curry, J.R. Hindley, and J.P. Seldin, *Combinatory Logic II* (Ibid.: 1972).

Notes to Chapter Fourteen

1. *Parts and Wholes*, Vol. I (Forskningsrådsmåmnden, Stockholm: 1983), pp. 123-128.

2. *Ibid.*, pp. 75-89.

3. In the *Logical Investigations* (Routledge and Kegan Paul, London: 1970), translated by J.N. Findlay, Vol. II, Investigation III.

4. Recall especially *Event, Reference, and Logical Form* and *Semiotics and Linguistic Structure*.

5. Recall "A Prolegomenon to Husserl's *Prolegomena*" in *Mind, Modality, Meaning, and Method*, and "On the Concept of Truth in Husserl's *Logische Untersuchungen*" in *Metaphysical Foundations: Mereology and Metalogic*.

6. Recall *Semiotics and Linguistic Structure*, Chapters V and VI, and III above.

7. *Algebra und Logik der Relative* (Teubner, Leipzig: 1895).

8. *Collected Papers*, Vol. III, XIII.

9. Recall "The Very Idea of a Logical From," in *Semiotics and Linguistic Structure*.

10. J.L. Austin, *How to Do Things with Words* (Clarendon Press, Oxford: 1962), p. 224.

11. The last three paragraphs are borrowed from *Semiotics and Linguistic Structure*, p. 238, with the kind permission of the publisher.

Notes to Chapter Fifteen

1. "Evolutionary Biology, Physiological Psychology, Process Philosophy," presented before the meeting of the Society for the Study of Process Philosophy in Evanston, Illinois, on March 15, 1984.

2. *Op.cit.*, pp. 322-323.

3. *Biology and Language* (Cambridge University Press, Cambridge: 1952).

4. Woodger spells this out in terms of R_{po}. Cf. *Principia Mathematics*, *91.

5. *Abstracts of the 7th International Congress of Logic, Methodology, and Philosophy of Science* (J. Huttegger OHG, Salzburg: 1983), Vol. 4, pp. 369-372.

6. Also presented at the Salzburg congress, *op.cit.*, pp. 302-372.

7. Recall the author's "Truth and Its Illicit Surrogates" in *Pragmatics, Truth, and Language*, and the remarks on Hintikka in "On Weeding the Garden of Semantics" in *Metaphysical Foundations: Mereology and Metalogic*.

8. Recall I, II and III above.
9. See especially David L. Hull, "Individuality and Selection," *Annual Reviews of Ecological Systems*, 11 (1980): 311-332.
10. Recall "On Being quā Being, Protometaphysics, and Epistemic Modality," in *Mind, Modality, Meaning, and Method*.
11. "Social Reality and Social Relations," *The Review of Metaphysics* XXXVII (1983): 3-20.
12. This paragraph is borrowed, with kind permission of the publisher, from the author's "On the Metaphysics of Social Reality" in *Metaphysical Foundations: Mereology and Metalogic*.
13. "Some Proposals Concerning Parts and Wholes," *Op.cit.*, pp. 115-122.
14. "Further Proposals to the Workshop on Parts and Wholes," to appear.
15. See his "Quantum Physics and the Philosophy of Whitehead," in *Boston Studies in the Philosophy of Science* (Humanities Press, New York: 1965), Vol. II, pp. 307-330.
16. "O Podstawach Matematiky," *Przeglad Filozoficzny*, 30-34 (1927-1931).
17. "The Calculus of Individuals and Its Uses," *The Journal of Symbolic Logic*, 5 (1940): 45-55.
18. Recall "An Approximative Logical Structure for Whitehead's Categoreal Scheme," in *Whitehead's Categoreal Scheme and Other Papers*.
19. In his *Understanding Whitehead* (The Johns Hopkins Press, Baltimore: 1962), p. 291.
20. A.N. Whitehead, *Essays in Science and Philosophy* (Rider and Co., London: 1948), p. 94.

Notes to Chapter Sixteen

1. In a lecture given at Harvard University on 9 March 1984.
2. *The Review of Metaphysics* XI (1957): 12-27.
3. Recall the appropriate definitions in the author's "On the Semantics of Sommer's 'Some *S*'," in *Mind, Modality, Meaning, and Method*.
4. Recall Chapters III and IV of *Mind, Modality, Meaning, and Method*, and I-III above.
5. To simplify, we omit all considerations concerning tense.
6. On virtual classes, recall "On the Philosophic Import of Virtual Classes" in *Belief, Existence, and Meaning*, and Chapters I and II n *Semiotics and Linguistic Structure*.
7. Recall *Truth and Denotation*, Chapters IV, V, and XI, and *Semiotics and Linguistic Structure*, Chapter VI.
8. Recall "'And'" in *Pragmatics, Truth, and Language*.
9. Recall the comments about "propositions" in III above, in *Toward a Systematic Pragmatics*, and also in *Belief, Existence, and Meaning*, Chapter II.
10. Recall I-III above.

11. Recall "A Close Look at Goodman's Symbol Systems," in *Mind, Modality, Meaning, and Method*.
12. Cf. George Hemphill, *op.cit.*
13. Recall "On Some Prepositional Relations," in *Semiotics and Linguistic Structure*.
14. As in the Hilbert-Ackermann *Principles of Mathematical Logic*, any edition.
15. Cf. *inter alia*, Tarski, *op.cit.*, and the author's *Truth and Denotation,* and *Semiotics and Linguistic Structure*.
16. *Mind* XXXVII (1928): 338-348.
17. Recall II above, *Toward a Systematic Pragmatics*, and *Belief, Existence, and Meaning*, Chapter III.
18. Recall "Some Comments on the *Insolubilia* and Gödel's Proof," in *Metaphysical Foundations: Mereology and Metalogic*.
19. *The Monist* XXXVIII (1928): 536-548.
20. *The Monist* XXXIX (1929): 281-319 and 440-472.
21. Recall also "On Epimathematics and Formal Meta-Systems" and "Toward an Objectual Dynamics of Logical Systems," in *Logico-Linguistic Papers*.
22. *Erkenntnis* 2 (1931): 242-261.
23. See W.V. Quine, *The Ways of Paradox and Other Essays* (Random House, New York: 1966), pp. 156-175 and the author's *Belief, Existence, and Meaning*, pp. 44 ff.
24. *The Monist* XLII (1932): 112-154.
25. Cf. John Kemeny, "Type-Theory vs. Set-Theory," abstract in *The Journal of Symbolic Logic* 15 (1950): 78.
26. Recall the discussion of Benson Mates in "On Mereology and the Heroic Course," in *Metaphysical Foundations: Mereology and Metalogic*. But see also XIX below.
27. *The Journal of Philosophy* XXX (1933): 288-293.
28. *The Philosophical Review* XLII (1933): 520-525.
29. C.I. Lewis, "Paul Weiss on Alternative Logics," *The Philosophical Review* XLIII (1933): 70-71.
30. As in *Meaning and Necessity*, 2nd ed., Appendix B.
31. Recall "The One and the Many: Logic_1 and Logics_2," in *Mind, Modality, Meaning, and Method*.
32. *Proceedings of the Seventh International Congress of Philosophy*, ed. by Gilbert Ryle (Oxford University Press, London: 1931), pp. 142-150.

Notes to Chapter Seventeen

1. In *American Philosophy: from Edwards to Quine*, ed. by R.W. Shahan and K.R. Merrill (University of Oklahoma Press, Norman: 1977), pp. 176-196.
2. Recall the paper by Joseph Earley discussed in XV above, and the literature cited therein.
3. Recall "A Homogeneous System for Formal Logic," referred to in V above.
4. We leave aside here the well-known Hilbert-Bernays way of introducing identity within a formal system.
5. *Journal of Philosophy* LXIV (1967): 5-22.
6. In *Belief, Existence, and Meaning*, pp. 51ff.
7. As in his *The Logical Syntax of Language*, p. 45, *Introduction to Symbolic Logic and Its Applications*, pp. 161 ff., and *Logical Foundations of Probability*, pp. 62 f.
8. Percy Bridgman, *The Nature of Physical Theory* (Princeton University Press, Princeton: 1936), pp. 59-60.
9. Recall "Toward an Objectual Dynamics of Logical Systems," in *Logico-Linguistic Papers*.
10. In *Metaphysical Foundations: Mereology and Metalogic*.
11. Recall "The Logic of Idealism and the Neglected Argument" in *Peirce's Logic of Relations and Other Studies*, and "Toward a Constructive Idealism" in *Primordiality, Science, and Value*.
12. Recall "On Metatheory Formulation and the World in the Making," in *Metaphysical Foundations: Mereology and Metalogic*.
13. Recall "On the Language of Metaphysical Theology: Hartshorne and Quine," also in *Metaphysical Foundations: Mereology and Metalogic*.
14. Cf. J.H. Wheeler, "Bits, Quanta, Meaning," to appear in the Caianiello Celebration volume, 1984.
15. D. Føllesdal, "Meaning and Experience," in *Mind and Language*, ed. by S. Guttenplan (Clarendon Press, Oxford: 1975), p. 254.
16. "Genesis and Observership," in *Foundational Problems in the Special Sciences*, ed. by R. Butts and J. Hintikka (D. Reidel Publ. Co., Dordrecht: 1977), pp. 3-33.
17. Recall "On Dialogism and Perception: Rivetti-Barbò" in *Logico-Linguistic Papers*.
18. On 'Under' recall the material of I-III above.
19. Eugene Wigner, *Symmetrics and Reflections* (Indiana University Press, Bloomington and London: 1967), p. 172.
20. Quoted in Wigner, *loc.cit.*

Notes to Chapter Eighteen

1. The formal development here is akin to that of the system S(P,OS) in Chapter XVII of *Metaphysical Foundations: Mereology and Metalogic* and is in harmony with some suggestions made in Chapter XVI of *Primordiality, Science, and Value*.
2. On the null individual, recall "Of Time and the Null Individual," in *Pragmatics, Truth, and Reference*.
3. As, e.g., in Church's *Introduction to Mathematical Logic* (Princeton University Press, Princeton: 1956), Chapters II and III.
4. As in Quine's *Mathematical Logic*, Chapter II.
5. As in Quine's "Logic Based on Inclusion and Abstraction," in *Selected Logic Papers* (Random House, New York: 1966).
6. As in D. Hilbert and P. Bernays, *Grundlagen der Mathematik*, Vol. II (Julius Springer, Berlin: 1939), §1.
7. H.W. Fowler, *A Dictionary of Modern English Usage* (Oxford University Press, New York and Oxford: 1965).
8. *Op.cit.*, p. 244.
9. A few examples of such are given in *Semiotics and Linguistic Structure*, pp. 173, 176, and 200 ff.

Notes to Chapter Nineteen

1. "The Unreasonable Effectiveness of Mathematics in the Natural Sciences," in *Symmetries and Reflections*, pp. 222-237.
2. As discussed, for example, in XIII above.
3. Recall *Principia Mathematica*, Part V, on the general theory of serial relations.
4. "On Mathematics and the Good," in *Essays in Science and Philosophy*, pp. 75-86.
5. Recall "Toward an Objectual Dynamics of Logical Systems," in *Logico-Linguistic Papers*.
6. Recall "Of Spiders and Bees: One Logic, not Two," in *Primordiality, Science, and Value*.
7. See his "Indication, Classes, Numbers, Validation," in *Essays in Science and Philosophy*, pp. 227-240.
8. (Harvard University Press, Cambridge: 1937).
9. A. Tarski, "The Semantic Conception of Truth," *Philosophy and Phenomenological Research* 4 (1944), pp. 367-368.
10. Recall "The One and the Many: Logic_1 and Logics_2," in *Mind, Modality, Meaning, and Method*.
11. *Logical Foundations of Probability*, pp. 8ff.

Notes to Chapter Twenty

1. Recall V above and "On Mereology and the Heroic Course," in *Metaphysical Foundations: Mereology and Metalogic*. Cf. also the Appendix below.
2. Recall especially *Events, Reference, and Logical Form; Semiotics and Linguistic Structure; Pragmatics, Truth, and Language*; and *Logico-Linguistic Papers*.
3. For contextual definition of '(*x* ord *y*)', see *D10a*, etc., in the Appendix below.
4. Recall, especially, *Semiotics and Linguistic Structure*, Chapters X-XII.
5. If an agent were always available we could write the definiendum here as '$p\ \mathrm{P}_e x,y$', expressing that person *p* takes *x* to be a part of *y* in act *e*, defined as '(<P>*e* . Act *e* . *e* $\mathrm{By}_{\mathrm{Agent}}\ p$. *e* $\mathrm{On}_{\mathrm{Object}}\ x$. *e* WithRespectTo *y*)'.
6. In "Bits, Quanta, Meaning."

In the FOUNDATIONS OF SEMIOTICS series the following volumes have been published and will be published during 1992:

1. PEIRCE, Charles S. (ed.): *STUDIES IN LOGIC by Members of the Johns Hopkins University* (1883). With an introd. by Max H. Fisch. Amsterdam, 1983.
2. WELBY, Victoria Lady: *What is Meaning? Studies in the Development of Significance.* (1903). With an introd. essay by Gerrit Mannoury. Amsterdam, 1983.
3. GÄTSCHENBERGER, Richard: *Grundzüge einer Psychologie des Zeichens* (1901). With an English summary. Amsterdam, 1987.
4. HOFFBAUER, Johannes Christoph: *Tentamina Semiologica, sive quaedam generalem theoriam signorum spectantia* (1789). Repr. of the orig. Latin text with an English transl. and introd. by Robert Innis. Amsterdam/Philadelphia, 1991.
5. WELBY, Victoria Lady: *Significs and Language.* Repr. with an introd. by H.W. Schmitz. Amsterdam, 1985.
6. WILKINS, John: *Mercury: or the Secret and Swift Messenger etc.* (1707). Tog. with an abstract of Dr. Wilkin's *Essay towards a Real Character and a Philosophical Language.* With an introd. essay. Amsterdam, 1984.
7. ESCHBACH, Achim & Jürgen TRABANT (eds): *HISTORY OF SEMIOTICS. Proceedings of the working group on the history of semiotics held at the 3rd Conf. of the German Society of Semiotics, Hamburg 1981.* Amsterdam, 1983.
8. IKEGAMI, Yoshihiko (ed.): *THE EMPIRE OF SIGNS. Semiotic Essays on Japanese Culture.* Amsterdam/Philadelphia, 1991.
9. PHARIES, David A.: *Charles S. Peirce and the Linguistic Sign.* Amsterdam, 1985.
10. DASCAL, Marcelo: *Leibniz. Language, Signs and Thought.* Amsterdam, 1987.
11. EKSTEIN, Rudolf: *The Language of Psychotherapy.* Amsterdam/Philadelphia, 1989.
12. BUSSE, Winfried & Jürgen TRABANT (eds): *LES IDEOLOGUES. Sémiotique, philosophie du langage et linguistique pendant la Révolution française.* Amsterdam, 1986.
13. FÓNAGY, Ivan: *Languages within Language. An evolutive approach.* Amsterdam/Philadelphia, n.y.p.
14. DELEDALLE, Gérard: *Charles S. Peirce, phénoménologue et sémioticien.* Amsterdam, 1987.
15. MORRIS, Charles W.: *Symbolism and Reality. A Study in the Nature of Mind.* Amsterdam/Philadelphia, n.y.p.
16. MARTIN, Richard M. : *LOGICAL SEMIOTICS AND MEREOLOGY.* Amsterdam/Philadelphia, 1992.
17. KEVELSON, Roberta: *Charles S. Peirce's Method of Methods.* Amsterdam, 1987.
18. DELEDALLE, Gérard (ed.): *SEMIOTICS AND PRAGMATICS.* Amsterdam/Philadelphia, 1989.
19. KOBERNICK, Mark: *Semiotics of the Drama and the Style of Eugene O'Neill.* Amsterdam/Philadelphia, 1989.
20. TOBIN, Yishai (ed.): *FROM SIGN TO TEXT: A SEMIOTIC VIEW OF COMMUNICATION.* Amsterdam/Philadelphia, 1989.
21. ECO, Umberto & Costantino MARMO (eds): *ON THE MEDIEVAL THEORY OF SIGNS.* Amsterdam/Philadelphia, 1989.
22. APHEK, Edna & Yishai TOBIN: *The Semiotics of Fortune-Telling.* Amsterdam/Philadelphia, 1989.
23. SCHMITZ, H. Walter (ed.): *ESSAYS ON SIGNIFICS. Papers presented on the occasion of the 150th anniversary of the birth of Victoria Lady Welby (1837-1912).* Amsterdam/Philadelphia, 1990.
24. MARTY, Robert: *L'Algèbre des signes.* Amsterdam/Philadelphia, 1990.
25. BÜHLER, Karl: *Theory of Language. The Representational Function of Language.* Amsterdam/Philadelphia, 1990.